T0342686

oxford **maths**
for australian schools

4

contents

NUMBER AND ALGEBRA

MEASUREMENT AND SPACE

STATISTICS AND PROBABILITY

OXFORD
UNIVERSITY PRESS

Practice

1 Write these numbers in words.

a   60 090 _____

b   48 723 _____

c   81 004 _____

2 Write the numerals for these numbers.

a   Thirty-eight thousand, four hundred and seven _____

b   Seventy-two thousand and ninety-five _____

c   Fifty-six thousand, five hundred and forty-nine _____

3 Expand these numbers by place value.

a   27 652 = _____ + _____ + _____ + _____

   + _____

b   48 075 = _____ + _____ + _____ + _____

   + _____

c   73 004 = _____

d   10 899 = _____

4 Use place value to find the answers.

a   40 000 + 3000 + 600 + 70 + 9 = _____

b   80 000 + 500 + 70 = _____

c   10 000 + 5 = _____

d   60 000 + 40 + 600 + 4000 + 6 = _____

Oxford University Press

Challenge

1 Circle the value of the **bold** digit.

a **7**3 471 | 3 ten thousands | 3 thousands | 3 hundreds

b 40**6**52 | 6 ten thousands | 6 thousands | 6 hundreds

c **9**8 403 | 9 ten thousands | 9 thousands | 9 hundreds

d 5**7**687 | 70 000 | 7000 | 700 | 70 | 7

e 64 0**9**3 | 90 000 | 9000 | 900 | 90 | 9

f **1**5 236 | 10 000 | 1000 | 100 | 10 | 1

---

2 Write the value of the **bold** digit.

a 4800**6** _____

b 9**1**728 _____

c **7**6 839 _____

d **2**3 064 _____

e 800**1**7 _____

f 35**5**55 _____

---

3 Circle the digit with the greater value in each number.

a 7 or 3? | 43 657

b 1 or 9? | 49 176

c 6 or 2? | 62 471

d 4 or 8? | 24 803

e 5 or 4? | 24 005

f 5 or 6? | 26 005

Is 9 always greater than 1?

1 384 can be renamed as:

300 + 80 + 4, 380 + 4 and 300 + 84

List as many ways as you can to rename 43 875.

2 The number for the highest attendance record for a cricket match at the Melbourne Cricket Ground uses the digits 0, 1, 3, 3 and 9.

a Write five different possibilities for the crowd numbers.

b The actual number is larger than 93 000 but smaller than 93 100. What two numbers could it be? _____

c The digit in the tens place of the attendance record is smaller than the digit in the ones place. What is the number? _____

3 Circle the correct answer.

a Who am I?

- I have a 7 in the thousands column.
- My tens column is 20 more than 60.
- I am larger than 36 000.
- I am smaller than 45 000.

I am:

37 489   47 489

34 789   37 469

b Who am I?

- The digit in my hundreds column is larger than the digit in my tens of thousands column.
- My tens column has a 0 in it.
- My ones column has an even number in it.

I am:

82 309   20 532

10 502   78 907

Oxford University Press

Practice

1 Is each number odd or even?

Don't forget to check the last digit!

a 60 090 _____

b 28 647 _____

c 55 050 _____

d 66 060 _____

e 34 621 _____

f 34 126 _____

g 45 478 _____

h 18 743 _____

2 Write the answers. Then complete the rule. The first rule has been done for you.

a 430 + 89 = _____

324 + 213 = _____

Rule: even + odd = odd

b 8001 + 2156 = _____

3205 + 1472 = _____

Rule: _____ + _____ = _____

c 5349 + 2321 = _____

6723 + 245 = _____

Rule: _____ + _____ = _____

d 950 – 624 = _____

478 – 212 = _____

Rule: _____ – _____ = _____

e 1729 – 312 = _____

3957 – 142 = _____

Rule: _____ – _____ = _____

f 2686 – 313 = _____

8798 – 357 = _____

Rule: _____ – _____ = _____

g 8 × 3 = _____

6 × 5 = _____

Rule: _____ × _____ = _____

h 5 × 15 = _____

11 × 7 = _____

Rule: _____ × _____ = _____

Challenge

1 Predict whether the answer will be odd or even. Then complete the calculation to check if you were right.

a I predict the answer will be:

Odd	Even

7384 + 6878 = _____

The answer is: | Odd | Even |

b I predict the answer will be:

Odd	Even

9374 – 8647 = _____

The answer is: | Odd | Even |

c I predict the answer will be:

Odd	Even

65 × 78 = _____

The answer is: | Odd | Even |

d I predict the answer will be:

Odd	Even

4007 + 6925 = _____

The answer is: | Odd | Even |

e I predict the answer will be:

Odd	Even

6809 – 372 = _____

The answer is: | Odd | Even |

f I predict the answer will be:

Odd	Even

36 × 100 = _____

The answer is: | Odd | Even |

---

2 Find the answers. Then use your knowledge of odd and even numbers to complete each rule.

a Angela picked 7548 grapes in a day. 5324 of them were sent to the local market. How many did she have left? _____

Rule: | Odd | Even | – | Odd | Even | = | Odd | Even |

b Akio baked 23 batches of cookies. Each batch had 37 cookies. How many did he bake altogether? _____

Rule: | Odd | Even | × | Odd | Even | = | Odd | Even |

c Jen had $4643 in her bank account. She earned another $1254. How much money does she have now? _____

Rule: | Odd | Even | + | Odd | Even | = | Odd | Even |

Oxford University Press

1 Julia has an odd number of stamps in her collection. The five digits in the number of stamps she has are: 4, 7, 3, 8, 2. How many stamps might she have? List as many possibilities as you can.

2 In each problem, rearrange any red digits so that the answer is odd.

a
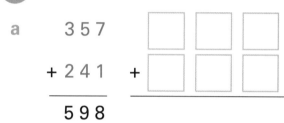

$$357 + 241 = 598$$

b
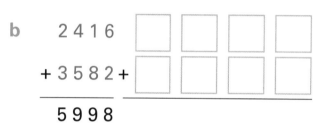

$$2416 + 3582 = 5998$$

c
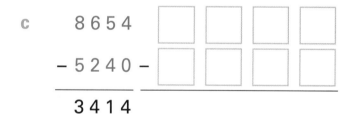

$$8654 - 5240 = 3414$$

d
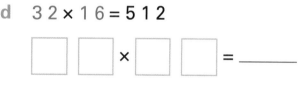

$$32 \times 16 = 512$$

3 In each problem, rearrange any red digits so that the answer is even.

a
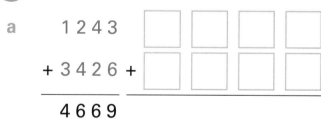

$$1243 + 3426 = 4669$$

b

$$6418 - 1243 = 5175$$

c
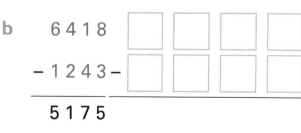

$$7065 - 3432 = 3633$$

d
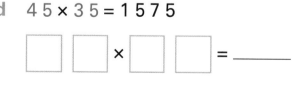

$$45 \times 35 = 1575$$

Practice

1 Devon was practising mental addition using three different strategies:

- rearranging numbers
- the jump strategy
- the split strategy.

Identify which strategy he used for each calculation.

a $537 + 148 = 537 \xrightarrow{+100} 637 \xrightarrow{+40} 677 \xrightarrow{+8} 685$ _____

b $67 + 32 + 18 + 23 = 90 + 50 = 140$ _____

c $692 + 247 = 800 + 130 + 9 = 939$ _____

d $151 + 65 + 319 + 135 = 470 + 200 = 670$ _____

e $2417 + 4281 = 6000 + 600 + 90 + 8 = 6698$ _____

2 Use a mental strategy to solve the addition problems. Record the strategy you used.

Strategy

a $473 + 214 =$ _____   _____

b $1246 + 632 =$ _____   _____

c $343 + 829 + 17 =$ _____   _____

d $4312 + 3654 =$ _____   _____

e $796 + 842 =$ _____   _____

f $314 + 89 + 111 + 226 =$ _____   _____

g $192 + 137 + 218 + 63 =$ _____   _____

Oxford University Press

1 This table shows different people's scores for an online game.

It can help to look for combinations that make the right total in the ones column.

Name	James	Sakura	Georgia	Josh	Fletcher	Jiang	Cara	Evie
Score	678	417	615	83	422	254	195	306

The scores of which two people together total:

a 500 _____ and _____

b 810 _____ and _____

c 1100 _____ and _____

d 560 _____ and _____

e 1293 _____ and _____

f 671 _____ and _____

g 278 _____ and _____

h 839? _____ and _____

2 The scores of which three people together total:

a 806 _____ , _____ and _____

b 643 _____ , _____ and _____

c 922 _____ , _____ and _____

d 1295 _____ , _____ and _____

e 1517? _____ , _____ and _____

3 The rounded scores of which two people have these approximate totals?

a 450 _____ and _____

b 700 _____ and _____

c 1040 _____ and _____

1 Zadie added together two 2-digit numbers made up of the digits 3, 4, 6 and 9. The answer was between 100 and 200. Use mental strategies to find as many possibilities as you can.

2 Zadie decided to add together two 3-digit numbers. The digits in the numbers were 1, 2, 5, 5, 7 and 8. If her answer was between 500 and 1000, what might the numbers have been?

3 Use the digits below to write a 4-digit addition problem that is easy to solve using the split strategy. Then solve your equation.

Digits: 0, 1, 3, 4, 5, 7, 8, 9

Oxford University Press

Practice

1 Use the split strategy, starting with the ones, to solve each addition problem. Then check your answer using vertical addition.

a   3846 + 2523

= _____

= _____

= _____

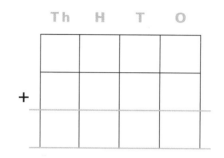

b   14 254 + 32 817

= _____

= _____

= _____

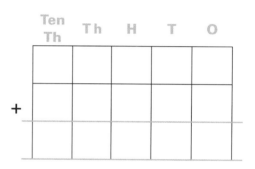

c   59 346 + 8282

= _____

= _____

= _____

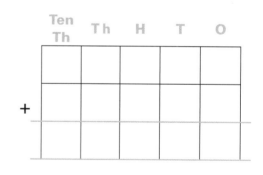

d   3794 + 26 436

= _____

= _____

= _____

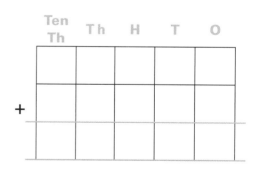

Challenge

Don't forget to align the place value columns if you are using vertical addition.

1 Use a written method of your choice to solve these problems. Show how you found the answer.

a Darcy and Seth's team scored 21 598 on Day 1 of competition and 58 701 on Day 2. What was their combined score?

b Isla and Maya's team scored 34 267 on Day 1 and 46 035 on Day 2. What was their total score?

c Whose team had the highest total score? _____

2 Mason and Owen competed against each other across three days. Use a written method of your choice to work out their total scores. Show how you found the answers.

Name	Day 1	Day 2	Day 3
Mason	48 071	6 153	31 249
Owen	25 641	35 879	23 950

a Mason's total

b Owen's total

c Who had the highest overall score? _____

d Who had the highest total after Day 2? _____

Oxford University Press

1 Mila's printer wasn't working properly when she printed her maths homework. Give as many suggestions as you can for what the missing numbers in each problem might be.

a

```
    ☐  7  8  ☐  6
+   ☐  3  1  ☐  5
─────────────────
    9  0  9  7  1
```

b

```
    3  2  ☐  0  7
+   5  1  ☐  8  ☐
─────────────────
    8  4  1  ☐  2
```

c

```
    4  ☐  ☐  7  4
+   3  5  ☐  2  5
─────────────────
    ☐  3  2  9  9
```

2

a Over two days, Archer sold 24 693 trading cards online. How many might he have sold on each day? Find at least three answers.

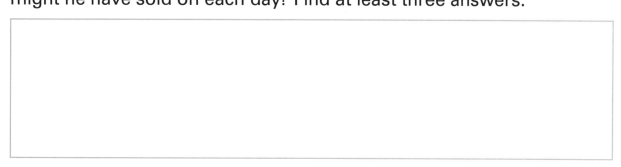

b If Archer sold the same number of cards over three days instead of two, how many might he have sold each day? Find at least three answers.

Practice

1 Round the numbers being subtracted to the nearest 10 or 100 to solve using the compensation strategy. Show your working on the empty number lines.

a 83 – 18 = _____

⟵—————————————————————————⟶

b 227 – 31 = _____

⟵—————————————————————————⟶

c 659 – 198 = _____

⟵—————————————————————————⟶

d 901 – 202 = _____

⟵—————————————————————————⟶

2 Circle the best mental strategy to use to find the difference: the compensation strategy or adding on. Then solve each problem.

a 731 – 724 = _____   Strategy: | Compensation | Adding on |

b 731 – 397 = _____   Strategy: | Compensation | Adding on |

c 1460 – 102 = _____   Strategy: | Compensation | Adding on |

d 1460 – 1448 = _____   Strategy: | Compensation | Adding on |

e 503 – 498 = _____   Strategy: | Compensation | Adding on |

f 3675 – 897 = _____   Strategy: | Compensation | Adding on |

Oxford University Press

Challenge

1 The Year 3 and 4 classes sold baked goods at the school fair. This table shows how many items each class sold.

Class	3A	3B	3C	3D	4A	4B	4C	4D
Items sold	298	487	748	314	604	1132	899	501

Use mental strategies to find the answers.

a How many more items did 4A sell than 3A? _____

b How many more items did 3D sell than 3A? _____

c Which two classes have a difference of 14 between the number of items they sold? _____

d Find the difference between the number of items sold by 4B and 4C.

_____

e Find the difference between the highest and lowest numbers of items sold. _____

- -

2 Estimate the answers by rounding the number being subtracted to the nearest 10 or 100. Then find the exact answer.

a 823 – 499   Estimated answer: _____ Exact answer: _____

b 3117 – 88   Estimated answer: _____ Exact answer: _____

c 758 – 149   Estimated answer: _____ Exact answer: _____

d 5431 – 602   Estimated answer: _____ Exact answer: _____

e 565 – 77   Estimated answer: _____ Exact answer: _____

f 3940 – 801   Estimated answer: _____ Exact answer: _____

g 932 – 468   Estimated answer: _____ Exact answer: _____

1 Jasper rounded the number he needed to subtract from 732 to the nearest 10 and got 582 as his answer. What might have been the original number he needed to subtract? Find at least three alternatives. Show your thinking.

2 Lexi rounded the number she needed to subtract from 3184 to the nearest 100 and got 2384. What might have been the original number she needed to subtract? Find at least three alternatives. Show your thinking.

3 Use the digits below to write a subtraction problem that is easy to solve using the compensation strategy. Then solve your equation.

1, 3, 5, 6, 8, 9

Oxford University Press

Practice

1 Use the split strategy, starting with the ones, to solve each subtraction problem. Then check your answer using vertical subtraction.

a 7526 – 3214   = _____ – _____ = _____

– _____ = _____

– _____ = _____

– _____ = _____

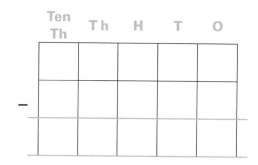

b 8140 – 5624   = _____ – _____ = _____

– _____ = _____

– _____ = _____

– _____ = _____

c 29 107 – 17 804 = _____ – _____ = _____

– _____ = _____

– _____ = _____

– _____ = _____

– _____ = _____

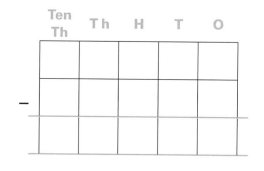

d 47 264 – 24 682 = _____ – _____ = _____

– _____ = _____

– _____ = _____

– _____ = _____

Challenge

1 Use a written method of your choice to solve the following. Show how you found the answer.

a Jedda had $34 317 in her bank account. She bought a car for $19 263. How much does she have left?

b Austin planned to travel 27 019 kilometres on his world trip. He has already travelled 22 849 kilometres. How far does he have left to go?

2 Caleb is aiming to raise $99 999 for charity. The table shows how much he has raised so far.

Month	Amount	Amount left to raise
January	$4725	
February	$11 092	
March	$8634	
April	$23 905	
May	$19 712	
June	$25 560	

Don't forget to align the place value columns if you are using vertical subtraction.

a Complete the final column for each month.

b Calculate the difference between the amounts in the month Caleb raised the most and the month he raised the least. _____

c Calculate the difference between the two largest amounts Caleb raised. _____

Oxford University Press

1 Water dripped on some of the numbers in Koko's homework book. Give as many suggestions as you can for what the missing numbers in each problem might be.

a
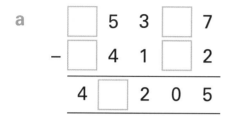

b
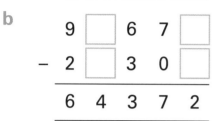

c
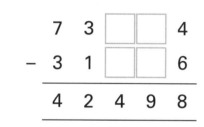

2

a Francesco had 43 705 grapes at the start of the day. If he had between 29 400 and 29 800 grapes at the end of the day, how many might he have sold? Show at least three solutions.

b If it took Francesco two days to sell the same number, how many grapes might he have sold each day? Show at least two solutions.

Practice

1 Write two division facts for each multiplication fact.

a $5 \times 4 = 20$ _____ ÷ _____ = _____   _____ ÷ _____ = _____

b $6 \times 7 = 42$ _____ ÷ _____ = _____   _____ ÷ _____ = _____

c $9 \times 8 = 72$ _____ ÷ _____ = _____   _____ ÷ _____ = _____

d $3 \times 7 = 21$ _____ ÷ _____ = _____   _____ ÷ _____ = _____

e $6 \times 9 = 54$ _____ ÷ _____ = _____   _____ ÷ _____ = _____

2 Write two multiplication facts for each division fact.

a $20 \div 10 = 2$ _____ × _____ = _____   _____ × _____ = _____

b $56 \div 8 = 7$ _____ × _____ = _____   _____ × _____ = _____

c $48 \div 6 = 8$ _____ × _____ = _____   _____ × _____ = _____

d $36 \div 4 = 9$ _____ × _____ = _____   _____ × _____ = _____

e $50 \div 5 = 10$ _____ × _____ = _____   _____ × _____ = _____

3 Write two multiplication and two division facts for each array.

a
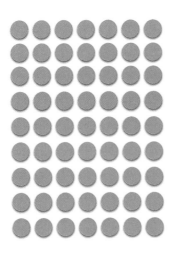

b

_____

_____

_____

_____

Oxford University Press

Challenge

Can you think of matching multiplication and division facts for each problem?

1 Use the following numbers to write two multiplication and two division facts.

a   3, 9, 27 _____

b   32, 4, 8 _____

c   10, 70, 7 _____

d   2, 18, 9 _____

e   72, 9, 8 _____

2 Double the 2s facts to find the 4s facts.

a 4×9

   $= 2 \times 9$ doubled

   $= 18$ doubled

   $=$ _____

b 4×7

   $= 2 \times 7$ doubled

   $=$ ____ doubled

   $=$ _____

c 4×8

   $= 2 \times 8$ doubled

   $=$ ____ doubled

   $=$ _____

3 Fill in the missing numbers.

a $6 \times \boxed{} = 54$

b $\boxed{} \times 9 = 90$

c $63 \div 9 = \boxed{}$

d $64 \div \boxed{} = 8$

e $\boxed{} \times 9 = 45$

f $16 \times \boxed{} = 160$

g $7 \times 5 = \boxed{}$

h $\boxed{} \div 7 = 4$

i $\boxed{} \div 9 = 9$

j $\boxed{} \times 10 = 1000$

k $8 \times \boxed{} = 24$

l $49 \div 7 = \boxed{}$

4   a   What is special about this list of numbers?   1, 4, 9, 16, 25

    _____

   b   What are the next two numbers in the list? _____

1 List as many multiplication and division facts as you can where the answer is 9.

2 Gianni bought 24 doughnuts to share with his friends. Write as many division facts as you can to show how he might have shared them.

3 Write a multiplication or division fact to solve each of the following.

How can you tell whether to use a division or multiplication fact?

a   Haley read eight pages every day for a week. How many pages did she read in total? _____

b   Ariel shared 63 toy cars equally into seven boxes. How many went into each box? _____

c   Lawson's factory has ten trucks to deliver 150 bikes. If each truck has the same number of bikes in it, how many are in each? _____

d   Michaela's five children each needed $7 for a school event. How much money did they need in total? _____

e   Stuart cut six watermelons into four pieces each. How many pieces did he have altogether? _____

Oxford University Press

Practice

1 Use extended multiplication to solve the following. Then rewrite each as a contracted multiplication to check your answers.

a   3 × 32

b   6 × 47

c   9 × 45

Extended

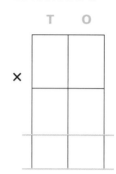

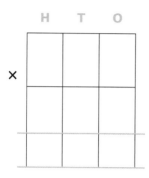

Contracted

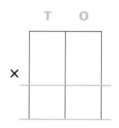

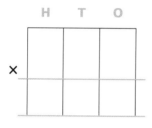

d   5 × 78

e   8 × 68

f   7 × 96

Extended

H   T   O

×

H   T   O

×

H   T   O

×

Contracted

H   T   O

×

H   T   O

×

H   T   O

×

1 Dustin was having trouble with written multiplication. Find the error in each problem. Then rewrite and solve it.

a

H	T	O
	8	3
×		6
	1	8
	4	8
	6	6

H	T	O
×		

b

H	T	O
	7	6
×		8
	4	8
5	6	0
5	0	8

H	T	O
×		

c

H	T	O
	9	0
×		8
	7	2

H	T	O
×		

d

H	T	O
	8	2
×		7
5	6	4

H	T	O
×		

2 Hugh's paper plane travelled 59 cm on his first throw. Use a written multiplication method to work out how far planes belonging to his classmates went.

a Luna's plane went four times further than Hugh's.

How can you check that your answers are correct?

b Kerry's plane went six times further than Hugh's.

c Jack's plane went seven times further than Hugh's.

d Hugh's second throw went nine times further than his first.

Oxford University Press

1 Use the digits 3, 6 and 8 to make at least three 1-digit by 2-digit multiplication problems. Then use your choice of methods to solve them.

2 Use the digits 4, 5, 7 and 9 to make at least five 1-digit by 2-digit or 1-digit by 3-digit multiplication problems. Then use your choice of methods to solve them.

Practice

1 Rewrite and solve each division problem. Then use contracted multiplication to check your answers.

a $84 \div 3$

b $95 \div 5$

c $99 \div 9$

```
      )
   T   O
  _____

× _____

  _____
```

```
      )
   T   O
  _____

× _____

  _____
```

```
      )
   T   O
  _____

× _____

  _____
```

---

2 Solve the 3-digit division problems.

a $3 \overline{)156}$

b $7 \overline{)189}$

c $8 \overline{)192}$

d $5 \overline{)205}$

e $4 \overline{)272}$

f $9 \overline{)243}$

Oxford University Press

Challenge

1 Kenji had to share some lollies equally between six people. Use a written division strategy to find which of the following can be equally divided into 6.

a 96

b 144

c 278

Divides equally?

Yes ☐ No ☐

Divides equally?

Yes ☐ No ☐

Divides equally?

Yes ☐ No ☐

2 Which numbers can also be equally divided by 8? Show your working.

a 96

b 144

c 278

Divides equally?

Yes ☐ No ☐

Divides equally?

Yes ☐ No ☐

Divides equally?

Yes ☐ No ☐

3 Elle was having trouble with division. Find the error in each problem. Then rewrite and solve it.

a
$$\begin{array}{r} 3\,1 \\ 4\,\overline{)\,1\,4\,4} \end{array}$$

b
$$\begin{array}{r} 1\,3 \\ 3\,\overline{)\,3\,0\,9} \end{array}$$

c
$$\begin{array}{r} 3\,9\,4 \\ 2\,\overline{)\,7\,9\,8} \end{array}$$

1 Hazel was collecting shells at the beach. She found between 160 and 260 shells. Imagine the number she found can be divided exactly by 9. How many might she have found? Show at least five options.

2 Imagine the number of shells Hazel found can **also** be divided exactly by 7. What might it be?

3 Tessa's dog chewed her homework. Find as many options as you can to show what the missing numbers might be.

a

b

c

Oxford University Press

Practice

1 whole			

$\frac{1}{2}$ $\frac{1}{2}$

$\frac{1}{3}$ $\frac{1}{3}$ $\frac{1}{3}$

$\frac{1}{4}$ $\frac{1}{4}$ $\frac{1}{4}$ $\frac{1}{4}$

$\frac{1}{5}$ $\frac{1}{5}$ $\frac{1}{5}$ $\frac{1}{5}$ $\frac{1}{5}$

$\frac{1}{6}$ $\frac{1}{6}$ $\frac{1}{6}$ $\frac{1}{6}$ $\frac{1}{6}$ $\frac{1}{6}$

$\frac{1}{8}$ $\frac{1}{8}$ $\frac{1}{8}$ $\frac{1}{8}$ $\frac{1}{8}$ $\frac{1}{8}$ $\frac{1}{8}$ $\frac{1}{8}$

$\frac{1}{10}$ $\frac{1}{10}$ $\frac{1}{10}$ $\frac{1}{10}$ $\frac{1}{10}$ $\frac{1}{10}$ $\frac{1}{10}$ $\frac{1}{10}$ $\frac{1}{10}$ $\frac{1}{10}$

$\frac{1}{12}$ $\frac{1}{12}$ $\frac{1}{12}$ $\frac{1}{12}$ $\frac{1}{12}$ $\frac{1}{12}$ $\frac{1}{12}$ $\frac{1}{12}$ $\frac{1}{12}$ $\frac{1}{12}$ $\frac{1}{12}$ $\frac{1}{12}$

1 Use the fraction wall to find how many:

a tenths are in $\frac{4}{5}$ _____

b sixths are in $\frac{1}{3}$ _____

c twelfths are in $\frac{1}{2}$ _____

d twelfths are in $\frac{3}{4}$ _____

e eighths are in $\frac{3}{4}$ _____

f fifths are in $\frac{4}{10}$. _____

- -

2 Draw lines to match each fraction with its equivalents.

$\boxed{\frac{2}{3}}$ $\boxed{\frac{5}{10}}$ $\boxed{\frac{2}{8}}$ $\boxed{\frac{4}{12}}$ $\boxed{\frac{3}{4}}$

$\frac{1}{4}$ $\frac{6}{8}$ $\frac{3}{6}$ $\frac{2}{6}$ $\frac{3}{12}$ $\frac{4}{8}$ $\frac{9}{12}$ $\frac{4}{6}$ $\frac{1}{2}$ $\frac{2}{4}$ $\frac{1}{3}$ $\frac{8}{12}$

Challenge

1 Dan and Zan like to share everything equally. Decide if they have equivalent fractions if:

a Dan has $\frac{6}{8}$ of a pizza and Zan has $\frac{8}{12}$.　| Equivalent | Not equivalent |

b Zan has $\frac{4}{10}$ of a cake and Dan has $\frac{2}{5}$.　| Equivalent | Not equivalent |

c Dan has $\frac{3}{4}$ of a glass of milk

 and Zan has $\frac{9}{12}$.　| Equivalent | Not equivalent |

d Zan has $\frac{5}{6}$ of a watermelon

 and Dan has $\frac{9}{10}$.　| Equivalent | Not equivalent |

2 The fractions below are not equivalent. Decide who has more in each case.

a Dan has $\frac{1}{2}$ a sandwich and Zan has $\frac{5}{8}$.　| Dan | Zan |

b Dan has $\frac{3}{13}$ of a muffin and Zan has $\frac{1}{3}$.　| Dan | Zan |

c Dan has $\frac{8}{10}$ of a pie and Zan has $\frac{3}{5}$.　| Dan | Zan |

d Dan has $\frac{1}{8}$ of a chocolate bar and Zan has $\frac{1}{4}$.　| Dan | Zan |

3 Draw and label an equivalent fraction for:

a $\frac{4}{5}$

b $\frac{9}{12}$

4 Draw and label two equivalent fractions for:

a $\frac{1}{2}$

b $\frac{4}{6}$

Oxford University Press

1. Herbie always eats halves, Carly eats quarters and Erwin eats eighths. If they share a cake, what fraction might each have? Draw and label your solutions.

2. Theo always eats thirds, Sam eats sixths and Toni eats twelfths. If they share a cake, what fraction might each have? Draw and label your solutions.

3. Caroline had two pieces of pizza. Added together, her share was $\frac{1}{2}$. What fraction might each piece have been?

Practice

1 Complete the table.

Diagram	Improper fraction	Mixed number
(three circles in quarters: two full, one with one quarter shaded)		$2\frac{1}{4}$
(rectangles in thirds)	$\frac{11}{3}$	
(circles in halves)	$\frac{11}{2}$	
(rectangles in fifths)		
(diamonds in quarters)		

2 Change the fractions to mixed or whole numbers and mark them on the number line.

a $\frac{12}{3} = $ _____

(number line 0 to 5)

b $\frac{9}{2} = $ _____

(number line 0 to 5)

c $\frac{15}{4} = $ _____

(number line 0 to 5)

3 Change the mixed numbers to improper fractions and mark them on the number line.

a $5\frac{1}{2} = $ _____

(number line labelled 0, $\frac{2}{2}$, $\frac{4}{2}$, $\frac{6}{2}$, $\frac{8}{2}$, $\frac{10}{2}$, $\frac{12}{2}$)

b $2\frac{1}{4} = $ _____

(number line labelled 0, $\frac{4}{4}$, $\frac{8}{4}$, $\frac{12}{4}$)

c $4\frac{2}{3} = $ _____

(number line labelled 0, $\frac{3}{3}$, $\frac{6}{3}$, $\frac{9}{3}$, $\frac{12}{3}$, $\frac{15}{3}$)

Oxford University Press

Challenge

1 Draw and label as both an improper fraction and a mixed number.

a $\frac{7}{2}$

b $\frac{10}{3}$

c $\frac{5}{3}$

d $\frac{13}{4}$

2

a Count by thirds from 1 to 4 using mixed numbers.

b Change the mixed numbers to improper fractions.

3 Count on to find the answers as both fractions and mixed or whole numbers.

a From Monday to Friday, Mai drank $8\frac{1}{4}$ litres of water. If she drinks another $\frac{5}{4}$ of a litre on Saturday, how many litres will she have had altogether? _____

b Dario used $3\frac{2}{3}$ litres of water on his pot plants on Monday. If he uses another $\frac{2}{3}$ of a litre on Tuesday, how many litres will he have used altogether? _____

c Shreya sold seven half pizzas on Saturday morning. If she sells another three halves, how many will she have sold altogether?

_____

1 Draw a number line to show counting by eighths from 1 to 3. Label it with both mixed numbers and fractions.

2 Ilan walked a whole number and $\frac{3}{8}$ of a kilometre home.

a Write three different mixed numbers to show how far he might have walked. _____

b Draw each of your mixed numbers on a number line.

3 The next day, Ilan's trip home was a whole number and $\frac{5}{6}$ of a kilometre.

a Write three different mixed numbers to show how far he might have walked. _____

b Draw each of your mixed numbers on a number line.

4 Shae sold cakes by the fifth. If she sold between six and eight cakes, list the amounts she might have sold as improper fractions.

Oxford University Press

Practice

1 Write each mixed number on the numeral expander.

a $2\frac{5}{10}$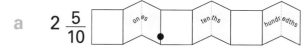

b $4\frac{38}{100}$

c $5\frac{9}{100}$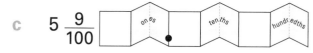

d $1\frac{91}{100}$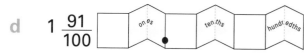

e $7\frac{23}{100}$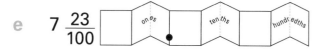

f $6\frac{6}{10}$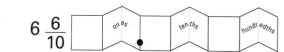

2 Write each decimal on the numeral expander.

a 0.6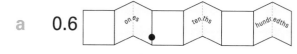

b 0.02

c 5.76

d 8.4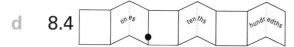

e 3.07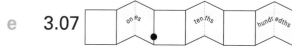

f 9.91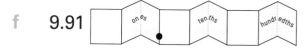

3 Mark each decimal in the correct place on the number line.

a 3.21

b 3.17

c 3.98

d 4.02

e 12.66

f 12.61

g 9.01

h 8.96

i 28.4

j 28.35

1 Write the decimal fraction that belongs on the number line at each arrow.

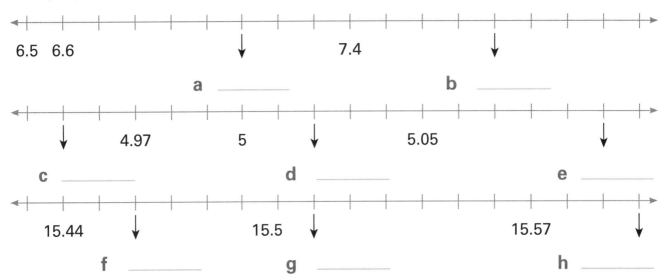

6.5 6.6   7.4

a _____   b _____

4.97   5   5.05

c _____   d _____   e _____

15.44   15.5   15.57

f _____   g _____   h _____

---

2 A Year 4 class competed in an online competition. Here are the top six scores:

Name	Abigail	Brent	Tia	Harley	Toby	Kimiko
Score	8.88	$8\frac{4}{10}$	$8\frac{24}{100}$	8.09	$8\frac{9}{100}$	$8\frac{9}{10}$

a Who had the highest score? _____

b Which two students had the same score? _____ _____

c Write Brent's score as a decimal. _____

d Write Abigail's score as a mixed number. _____

e How much higher was Abigail's score than Tia's? _____

---

3 Use the numbers in the box to answer the questions.

50.25	50.63	49.10	49.09	50.8	$49\frac{9}{10}$	$50\frac{12}{100}$	$49\frac{99}{100}$	$49\frac{50}{100}$	$50\frac{3}{100}$

a Which number is closest to 50? _____

b Which number is closest to 49? _____

c Which number is closest to 51? _____

d Which two numbers have a zero in the tenths column when written as a decimal? _____ _____

Oxford University Press

1 April ran 100 metres and timed herself to 2 decimal places.

    a If the digits in her time were 3, 6, 2 and 4, what might her time have been? List five different answers.

       _____   _____   _____   _____   _____

    b From your answers, which is the fastest time? _____

    c From your answers, which is the slowest time? _____

2 Fill in the missing digits so the numbers go from smallest to largest.

23. ☐ 4 ,   23. 6 ☐ ,   23. ☐ 5 ,   23. ☐ 1 ,   2 ☐ .91 ,   24.9 ☐

3   **a** Colour the hundred grid using five different colours. Choose a different number of squares for each colour.

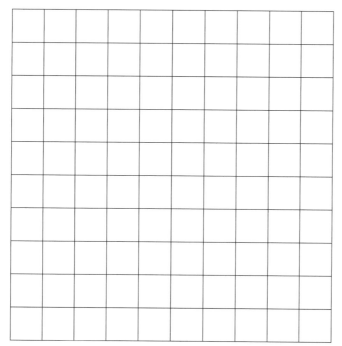

    b Complete the table to show the number of squares you have of each colour as a common fraction and a decimal fraction of the whole.

Colour	Fraction	Decimal

Practice

1 Round the price of each item to the nearest 5 cents.

a $8.78

Rounds to _____

b  $56.24

Rounds to _____

c  $89.87

Rounds to _____

d $0.98

Rounds to _____

e  $19.99

Rounds to _____

f $132.53

Rounds to _____

g $518.61

Rounds to

_____

h  $3.02

Rounds to _____

2 Round each amount in the table to the nearest 5 cents. Then show how much change you would get from $10, $20 and $50.

What do you notice about the change amounts?

Amount	Rounds to	Change from $10	Change from $20	Change from $50
$4.32				
$0.06				
$3.74				
$4.98				
$1.17				

Oxford University Press

Challenge

1 This table shows the price of five items Jeb bought. You may use a calculator to help you answer the questions.

Item	Book	Shirt	Jeans	Shoes	Cake
Cost	$37.41	$86.56	$52.97	$71.84	$28.03

a   Would the total cost of the items round up or down? _____

b   How much change would Jeb get if he paid with $300? _____

c   Which two items together would give the largest amount of change

from $100? _____   _____

d   How much change would Jeb get from $100? _____

e   Which two items together would give the smallest amount of

change from $200? _____   _____

f   How much change would Jeb get from $200? _____

- -

2 Amber has a budget of $150 to spend on presents for her family.

a   If she buys a hat for her brother for $18, how much does she have

left? _____

b   If she then spends $27 on a scarf for her mum, how much does

she have left? _____

c   If she then spends $48 on a sports bag for her dad, will she have
enough money to buy two books, which each cost $29.50, for her
cousins? _____

d   After spending all her money, Amber remembered that she needed
to buy a present for her grandparents. She earned money from
babysitting to pay for it. If she received $28.45 change, how
much might she have earned?

_____

e   Using the amount you think
Amber earned, how much did
the gift for her grandparents
cost? _____

> Calculate the change
> you should receive next
> time you go to the shops
> to make sure it's the
> right amount.

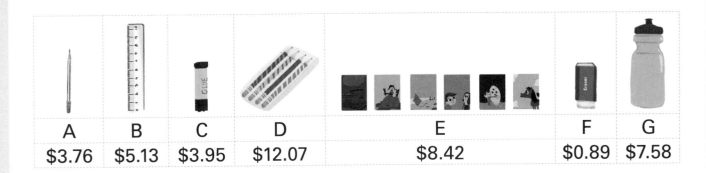

A	B	C	D	E	F	G
$3.76	$5.13	$3.95	$12.07	$8.42	$0.89	$7.58

1 Imagine you are buying pairs of these items and need to work out which pairs have a total price that would be rounded up.

a Record the letters and total price of each pair.

b Record the price you would round up to.

c Record the change you would get from $50 for each pair.

2 Kendrick bought three items from the shops. If the amount rounded to $12.50, what might the cost of each item be? List at least three solutions.

3 Amelia received $7.20 in change when she bought two items.

a What might the total cost of the items be?

b How much might each item have cost?

Oxford University Press

Practice

1 Write the rule and complete each number pattern.

a

1	2	4	8	16

Rule: _____

b

3	9	15	21

Rule: _____

c

152	148	144	140

Rule: _____

2 a Circle all the multiples of 3 on the number chart.

b Shade the multiples of 6.

c Which numbers are both
circled and shaded? _____

What happens in the tens column when you count by 9?

d Tick all the multiples of 9.

e How many ticked numbers are also shaded and circled? _____

f What would the pattern in the ones column be if you start
at 2 and count by 9s? _____

0	1	2	3	4	5	6	7	8	9
10	11	12	13	14	15	16	17	18	19
20	21	22	23	24	25	26	27	28	29
30	31	32	33	34	35	36	37	38	39
40	41	42	43	44	45	46	47	48	49
50	51	52	53	54	55	56	57	58	59
60	61	62	63	64	65	66	67	68	69
70	71	72	73	74	75	76	77	78	79
80	81	82	83	84	85	86	87	88	89
90	91	92	93	94	95	96	97	98	99

1 Fill in the missing numbers on each number pattern grid. Then write the rule for the pattern.

a

3	11	19	27	35	43
51		67	75	83	91
99	107	115		131	139
147	155		171	179	
195	203	211		227	235
	251		267	275	

Rule: _____

b

253	246	239	232		218
211		197	190	183	176
169	162	155		141	134
127	120	113	106	99	92
85	78	71	64		50
43		29	22	15	8

Rule: _____

2 Predict the calculator answer for these operations. Then use the calculator.

a $1 \times 3 =$. My prediction: _____ Actual answer: _____

b Press = again. My prediction: _____ Actual answer: _____

c Press = again. My prediction: _____ Actual answer: _____

d Rule: _____

3 Circle the numbers that are multiples of:

a	7	49	27	83	42	16	51	55	35	63	72
b	8	23	78	66	56	96	100	38	32	53	88
c	9	27	61	78	72	98	104	81	63	45	37

Oxford University Press

1 Spencer made a function machine with a rule of *multiply by 6*. What might the numbers in and out have been? Write at least five possibilities.

2 Spencer made another function machine.

a If he put in 4 and 12 came out, what might the rule be? Write two possibilities.

b If Spencer then put in 9 and 27 came out, what would the rule have to be? _____

c Write five more pairs of 'in' and 'out' numbers using the rule.

3 The last number in Matilda's multiplication number pattern was 60.

a What might the rule be? Find as many possibilities as you can.

b Choose one of your rules from **a**. If Matilda kept counting, what would the next 5 numbers be?

Rule: _____   Numbers: _____ _____ _____ _____

c Using the same rule, what would the five numbers before 60 have been?

_____ _____ _____ _____ _____

Practice

1 Find the matching pairs of equations and rewrite them as an equivalent number sentence, e.g. $5 + 3 = 10 - 2$.

| 52 + 21 | 63 ÷ 7 | 76 – 52 | 48 – 28 | 5 + 19 |

| 38 + 16 | 10 × 2 | 28 + 45 | 6 × 9 | 38 – 29 |

2 Fill the gaps to complete the equivalent number sentences. You may use a calculator.

a $17 + \boxed{} = 48 - 22$

b $80 \div 10 = \boxed{} - 36$

c $45 - 26 = 73 - \boxed{}$

d $9 \times 8 = 42 + \boxed{}$

e $38 + 12 = \boxed{} \div 2$

f $74 - \boxed{} = 11 \times 6$

g $94 - \boxed{} = 21 + 37$

h $\boxed{} \times 2 = 90 \div 9$

3 Circle all the equations that are equivalent to $48 + 37$. You may use a calculator.

| 100 – 15 | 3 × 25 | 47 + 38 | 19 + 56 | 170 ÷ 2 |

| 61 + 27 | 109 – 24 | 98 – 14 | 17 × 5 | 71 + 14 |

Oxford University Press

1 Insert the correct signs to make both sides of each equation balance.

a 58 ☐ 17 = 18 ☐ 23

b 27 ☐ 3 = 48 ☐ 39

c 32 ☐ 24 = 8 ☐ 7

d 95 ☐ 44 = 3 ☐ 48

e 12 ☐ 4 = 6 ☐ 8

f 35 ☐ 57 = 99 ☐ 7

g 86 ☐ 40 = 78 ☐ 32

h 8 ☐ 4 = 63 ☐ 31

2 Write a number sentence to solve each problem.

a Linus subtracted a number from 87 and found that the answer was the same as 31 + 28. What was the number?

b Hala had some cupcakes. After she made 24 more, she had twice as many as Hannah, who had 33. How many cupcakes did Hala start with?

c Jacob multiplied a number by 3 and found that the answer was the same as 92 − 47. What number did he start with?

d Andy used 21 sheets of paper and ended up with the same number as Ryan and Sharni combined. If Ryan had 34 sheets and Sharni had 27, how many did Andy start with?

1. Emily's little brother rubbed out some of the numbers in her homework. Record at least three options for what the missing numbers might be.

   $2\boxed{} + 3\boxed{} = \boxed{}9 - \boxed{}3$

2. Nicholas collected twice as many shells on one day as Trudy did on two days. Record number sentences to show how many shells they might have collected.

3. Alice sold between 30 and 110 scarves in two weeks. If she sold 32 more scarves in the second week than in the first week, how many scarves might she have sold each week, and how many might she have had to start with? Find at least three solutions and write them as number sentences.

Oxford University Press

Practice

1 Estimate, then measure, the length or height of each animal picture in millimetres.

a

Estimate: _____
Height: _____

b

Estimate: _____
Length: _____

c

Estimate: _____
Length: _____

d

Estimate: _____
Length: _____

e

Estimate: _____
Length: _____

f

Estimate: _____
Height: _____

2 Find the length of each line in centimetres.

a _____ Length: _____

b _____ Length: _____

c _____ Length: _____

d _____ Length: _____

e _____ Length: _____

3 Which item is longer?

a A 30 cm tie or a 250 mm tie _____

b A $17\frac{1}{2}$ cm worm or a 17 mm worm _____

c A 235 mm book or a 25 cm book _____

d A 2 m rope, a 20 cm rope or a 200 mm rope _____

1   Record the measurement shown on each ruler in centimetres.

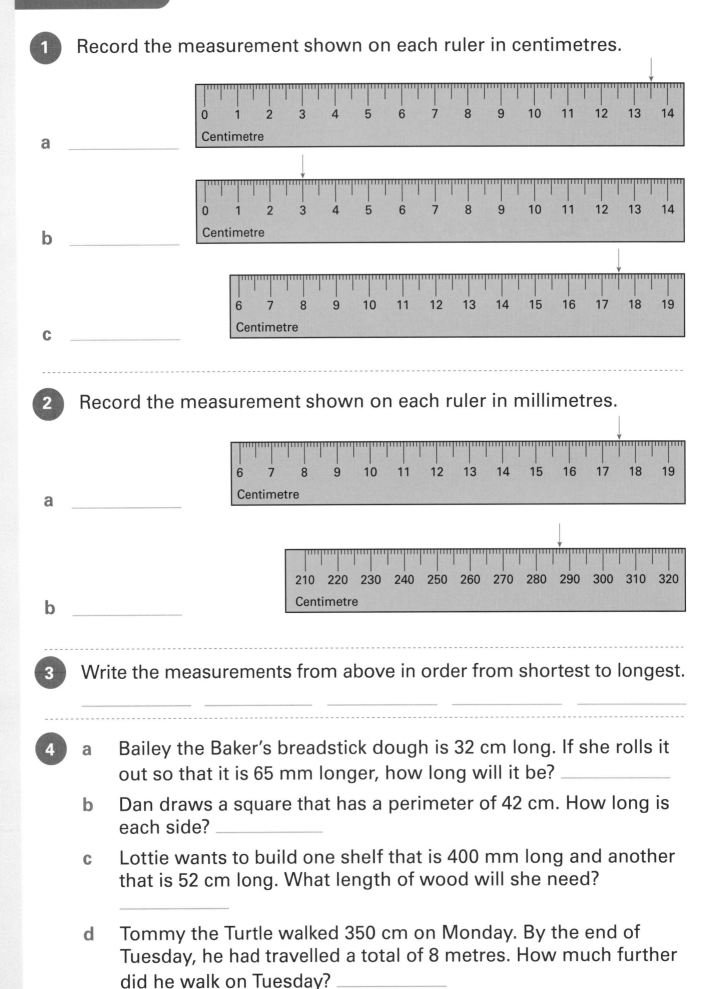

a   _____

b   _____

c   _____

2   Record the measurement shown on each ruler in millimetres.

a   _____

b   _____

3   Write the measurements from above in order from shortest to longest.

_____   _____   _____   _____   _____

4   a   Bailey the Baker's breadstick dough is 32 cm long. If she rolls it out so that it is 65 mm longer, how long will it be? _____

    b   Dan draws a square that has a perimeter of 42 cm. How long is each side? _____

    c   Lottie wants to build one shelf that is 400 mm long and another that is 52 cm long. What length of wood will she need?

        _____

    d   Tommy the Turtle walked 350 cm on Monday. By the end of Tuesday, he had travelled a total of 8 metres. How much further did he walk on Tuesday? _____

Oxford University Press

1 Draw:

a a four-sided shape with a perimeter of exactly 24 cm

b a shape with a perimeter of exactly 260 mm.

2 Nabeel drew a five-sided shape with a perimeter of 65 cm. What might the length of each side be? Find at least five solutions.

3 Melita has three lots of 270 m of fencing. Draw and label the sides of three different-shaped paddocks she could fence.

Practice

1   Write cm² or m² to show the best unit to measure the area of:

a   the school oval _____

b   your bedroom floor _____

c   the cover of this book _____

d   the top of your desk _____

e   the school car park _____

f   a credit card. _____

2   Record the area of each shape in cm². Imagine the grid is made up of 1 cm squares.

a

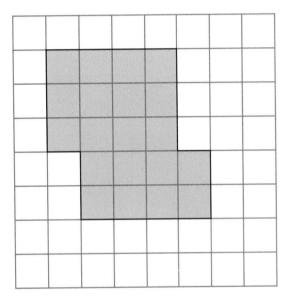

_____

b

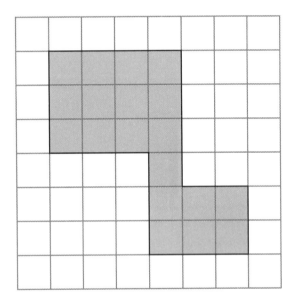

_____

c

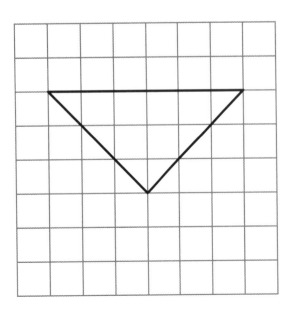

d

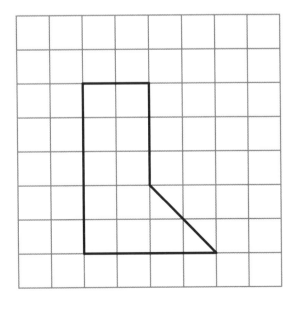

_____

Oxford University Press

How much bigger is the mobile phone than the biscuit?

1 Redraw each item on the grid to show its area. The first one has been done.

a 49 cm²   b  8 cm²   c 4 cm²

d 15 cm²   e  32 cm²   f 12 cm²

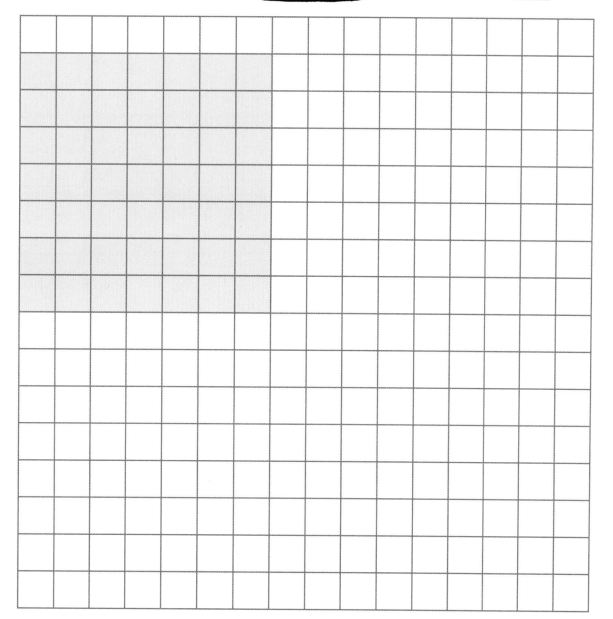

2 Look at the items on the grid.

a Write the letter of the item with the largest area. _____

b What is the combined area of the biscuit and the eraser? _____

1. Use your ruler to help you draw as many different rectangles as you can with an area of 12 cm². Label the lengths of the sides of each one.

2. Nate bought some floor tiles that were 1 m². Record the area of his room if it is:

   a   3 tiles long and 3 tiles wide _____

   b   5 tiles long and 4 tiles wide _____

   c   7 tiles long and 2 tiles wide _____

   d   8 tiles long and 3 tiles wide _____

   e   6 tiles long and 3 tiles wide _____

   f   9 tiles long and 4 tiles wide. _____

Oxford University Press

Practice

1 Write the volume of each object in cm³. Imagine each block is 1 cm³.

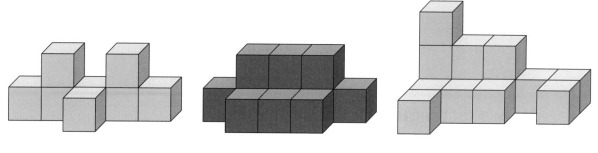

a _____ b _____ c _____

2 Draw an object with a volume of:

a 12 cm³

b 9 cm³.

3 Mark the capacity of each container on the jug.

a

b

c

d

Challenge

1   **a**   Draw a rectangular prism with three layers and 6 square centimetres in each layer.

     b   What is the volume of your prism? _____

---

2   **a**   Draw a rectangular prism with four layers and 4 square centimetres in each layer.

     b   What is the volume of your prism? _____

---

3   Write the amounts shown in each jug in millilitres, and in litres and millilitres.

a

_____   _____

b

_____   _____

c

_____   _____

d

_____   _____

e

_____   _____

f

_____   _____

Oxford University Press

1 Draw as many different rectangular prisms as you can with a volume of 24 cm³. For each object, write how many layers it has and how many cm³ are in each layer. Your prisms do not need to be drawn to scale.

2 Draw a measuring jug that you could use to accurately measure a smaller amount, such as 250 mL, and a larger amount, such as 4 L. Label the increments on your jug.

You might like to use a ruler to help you space the markings on your measuring jug.

Practice

1 Write the letters for the pictures in order from lightest to heaviest.

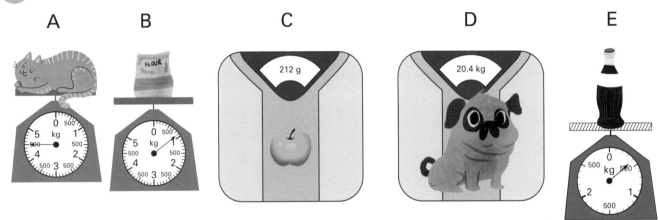

A   B   C   D   E

lightest _____ _____ _____ _____ _____ heaviest

---

2 Find and record the missing masses to balance the scales.

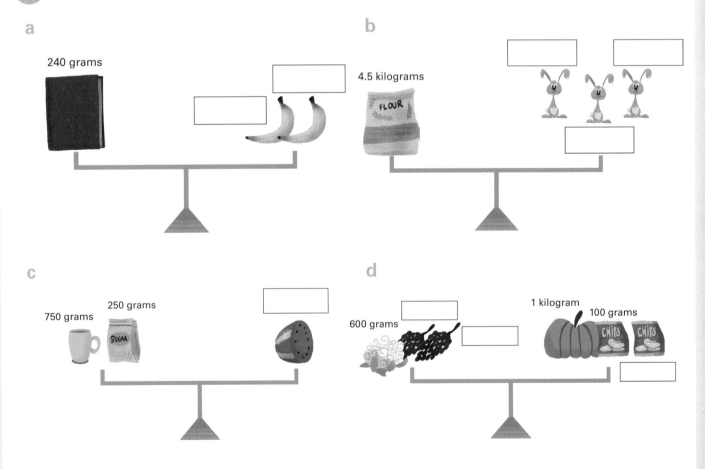

a

240 grams

b

4.5 kilograms

c

750 grams   250 grams

d

600 grams   1 kilogram   100 grams

Oxford University Press

1 Tully made some dough with 250 grams of flour, 3 eggs and 120 grams of sugar. If each egg had a mass of 55 grams, what was the total mass of the ingredients?

_____

2 Angela's dough had 220 grams of flour, 2 eggs, 150 grams of sugar and 130 grams of butter in it.

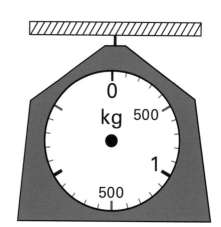

a Mark the total mass of Angela's ingredients on the scale.

b Does Angela's or Tully's dough have a greater mass? _____

c If one cup of flour is 250 grams, how many cups will Angela need if she uses 1 kilogram of flour? _____

3 The table below shows the ideal mass of different breeds of dogs. Mark the letter of each on the correct place on the scale.

	Breed	Ideal mass
A	Beagle	14 kg
B	Boston terrier	4.5 kg
C	Chihuahua	2.7 kg
D	Border collie	19 kg
E	Greyhound	29 kg
F	Toy poodle	3.5 kg
G	Pug	6.3 kg
H	Cattle dog	21 kg

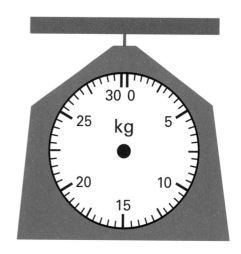

1 A 1 litre bottle of milk has a mass of about 1 kilogram.

a List five items with a mass of less than 1 kg and estimate what their masses are.

b List five items with a mass of more than 1 kg but less than 20 kg and estimate what their masses are.

_____

_____

_____

_____

_____

_____

_____

_____

_____

_____

2 Draw a scale and mark on it the mass of each of the ten items you listed in question 1.

Oxford University Press

Practice

1 Imagine the temperature today is 25°C. Decide whether the item or place in each picture is colder or hotter than this. Circle your choice.

a   hotter   colder

b   hotter   colder

c   hotter   colder

d   hotter   colder

2 Label the thermometer with the letter for each item at the correct temperature.

a   human body temperature   37°C

b   butter   2°C

c   hot soup   71°C

d   frozen water (ice)   0°C

e   room temperature   20°C

f   body temperature of a goat   40°C

g   highest daytime temperature
    recorded in Australia   49°C

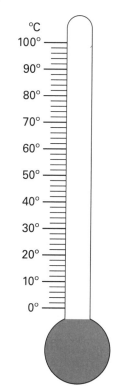

What can you think of that has a temperature of over 100°C?

1 The thermometer shows the temperature for each hour from 8 am to 8 pm on Planet Creeb.

a Complete the table to show the temperature at each time.

Time	Temperature
8 am	
9 am	
10 am	

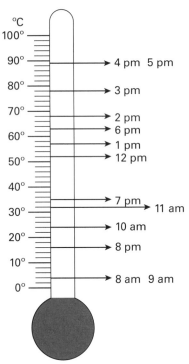

b At what time of day was it the hottest? _____

c At what time of day was it the coldest? _____

2 The thermometer shows hourly temperatures recorded in Alice Springs.

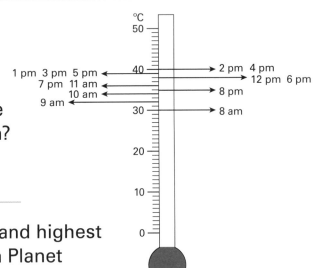

a Was the temperature higher in Alice Springs or on Planet Creeb at 11 am?

_____

b What about at 8 pm? _____

c Was the range between the lowest and highest temperatures for the day greater on Planet Creeb or in Alice Springs? _____

d At what time was the temperature in Alice Springs 10 degrees higher than on Planet Creeb? _____

e What season do you think it is in Alice Springs? Why?

_____

Oxford University Press

Mastery

1 Write the name of the place where you live and the top temperature for today.

_____

a List five places or items that are hotter than today's top temperature and estimate what their temperatures are.

_____

_____

_____

_____

_____

b List five places or items that are colder than today's top temperature and estimate what their temperatures are.

_____

_____

_____

_____

_____

2 Draw your own thermometer and mark on it the temperatures you estimated for each of the ten places and things that you listed in question 1.

Practice

1 Write the times shown on the clocks from earliest to latest.

pm   am   am

pm   pm   am

earliest _____ _____ _____ _____ _____ _____ latest

- -

2 The table below shows where Dave went during the day and when he left and arrived.

a Write his travel times in the last column.

Departed from	Time	Arrived at	Time	Travel time
Salt City	7:43 am	Butter Bay	7:58 am	
Butter Bay	8:16 am	Herb Hill	9:03 am	
Herb Hill	10:30 am	Garlic Green	11:09 am	
Garlic Green	11:14 am	Cherry Creek	11:23 am	
Cherry Creek	11:48 am	Banana Bend	12:07 pm	
Banana Bend	1:51 pm	Salt City	2:02 pm	

b How long did Dave spend at Cherry Creek? _____

c Which place did he spend the longest time at? _____

d Which place did he spend the shortest time at? _____

e How long was Dave away from Salt City? _____

f Did Dave get back to Salt City in the morning or the afternoon?

_____

g How many seconds did Dave spend in Garlic Green? _____

Oxford University Press

Challenge

1 Orla's school day starts at 8:45 am and ends at 3:15 pm.

a How long is the school day? _____

b If recess is 40 minutes and lunch is 50 minutes, how much time does Orla spend in class each day? _____

c How much time does she spend in class over a week?

_____

d Orla's Italian lesson is on Wednesday. It starts at 11:40 am and goes for 50 minutes. Show the finish time on the clock.

e Is the finish time an am or a pm time?

_____

f If Orla leaves school at 1:20 pm on Friday, how long did she spend there that day? _____

---

2 The clocks below show when Orla left home and when she arrived at school.

a Write the times next to the clocks. Then write as am or pm time.

b How long did the journey take? _____

c If Orla leaves school at 3:27 pm, what time will she arrive home?

_____

---

3 **a** Orla's birthday is in 123 days. About how many months is that?

_____

b It takes her 155 seconds to walk from her classroom to the library. How long is that in minutes and seconds? _____

c School holidays are in five weeks' time and the class excursion is in 51 days. Which will happen sooner? _____

1. Draw your own analog clock and show the time you got up this morning. Write the time as an am or pm time next to your clock.

2. If it takes you 33 minutes to get home from school, what time might you leave and what time would you get home? Record at least three different options.

3. List three things you do that take:

    a   about a minute _____

    b   about an hour _____

    c   about a month. _____

Oxford University Press

Practice

1 a Name the large shape. _____

b Is it regular or irregular? _____

c What two smaller shapes is it split into?

_____ _____

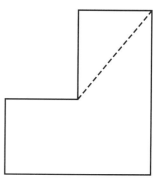

2 a Name the large shape. _____

b Is it regular or irregular? _____

c What two smaller shapes is it split into?

_____ _____

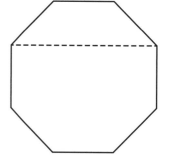

3 Which of the following shapes could you combine with a square to make:

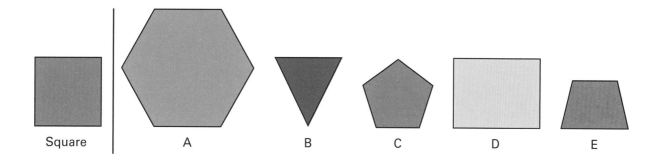

Square   A   B   C   D   E

a a pentagon _____ b a rectangle _____

c a hexagon _____ d an octagon? _____

4 What is the smallest number of triangles you could use to make:

a a square _____ b a trapezium _____

c a regular hexagon _____ d a regular pentagon? _____

1 **a** Circle the name of any 2D shapes that cannot be seen on the house.

hexagon   square   octagon

rectangle   kite

circle   trapezium   semi-circle

isosceles triangle   equilateral triangle

b Draw a building in which there are at least four different 2D shapes. Write the names of the shapes that you use.

These are the shapes I used:

_____

_____

_____

_____

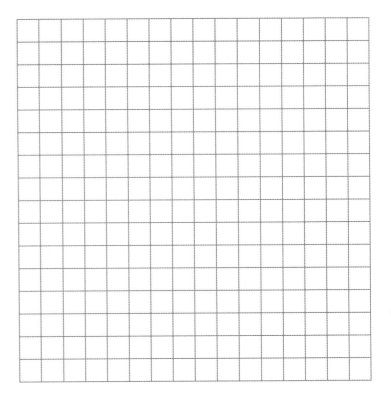

2 **a** Draw two irregular shapes on the grid.

b Write the area of each shape in cm². Imagine the grid is made up of 1 cm squares.

Shape 1: _____

Shape 2: _____

c Name each shape.

Shape 1: _____

Shape 2: _____

1

a   Draw as many different shapes as you can with an area of 12 cm². The shapes do not need to be drawn to scale.

b   Label each of your shapes with its name.

c   Choose one shape and draw a line to split it into two shapes.
Name the two shapes. _____ _____

d   Choose a different shape and split it into three shapes.
Name each shape. _____ _____ _____

2   Draw and name as many shapes as you can using three or more of the following shapes.

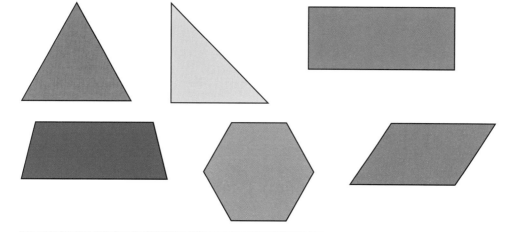

Practice

1   Draw the following objects.

   a   a rectangular prism

   b   a triangular pyramid

   c   a triangular prism

   d   a pentagonal pyramid

----

2   On each of your drawings in question 1, label:

   a   a corner

   b   an edge

   c   a face.

The side faces of prisms are rectangles.

Oxford University Press

Challenge

1 Match each object with its top view.

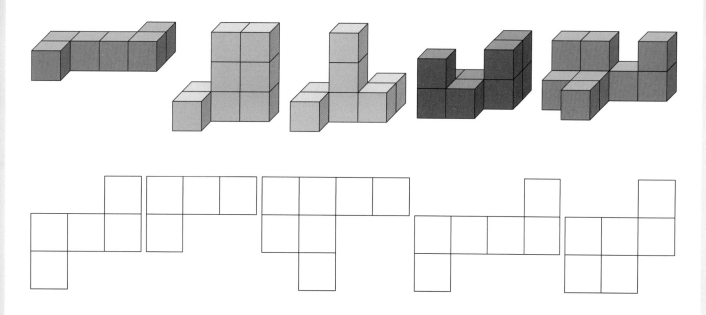

---

2 Record the number of faces, edges and corners for each object.

a

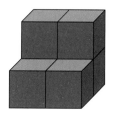

faces _____

edges _____

corners _____

b

faces _____

edges _____

corners _____

c

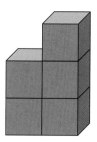

faces _____

edges _____

corners _____

d

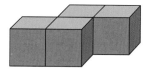

faces _____

edges _____

corners _____

1 Alida had six 2D shapes. She used them all to make a 3D object. What might the object have looked like? Draw and name at least two options.

2 Mario made an object that has a rectangle as its side view. What might it have looked like? Draw and name at least three options.

3 **a** If all the faces of Mario's object were rectangles, what would the object be? _____

b If his object had three rectangular faces and two faces of another shape, what would it be? _____

Oxford University Press

Practice

1 **a** Draw an angle that is smaller than a right angle.

b Label your angle with its angle type.

c Draw an angle that is larger than a right angle.

d Label your angle with its angle type.

---

2 Look at the picture.

a Circle five right angles in blue.

b Circle five angles smaller than a right angle in green.

c Circle five angles greater than a right angle in red.

3 Circle and label:

a a reflex angle

b an obtuse angle

c an acute angle.

---

4 Circle the shapes that have a right angle.

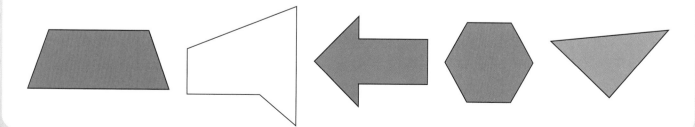

Challenge

1 Draw a shape that has exactly:

    a four right angles   **b** one right angle   **c** no right angles.

2
  a Draw a five-sided shape.

  b Label each angle with its angle type.

  c Draw a six-sided shape.

  d Label each angle with its angle type.

3 Which of the shapes below have:

  a only acute angles _____

  b at least one obtuse angle _____

  c a reflex angle? _____

A   B   C   D   E

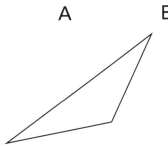
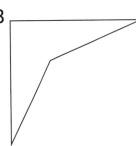
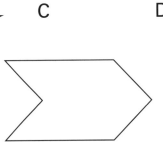
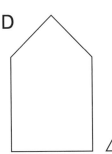
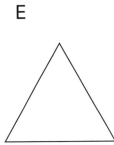

Oxford University Press

1 Draw your own picture that has:

- at least three right angles
- at least one reflex angle
- at least one revolution

Label one of each angle type.

- at least five acute angles
- at least one straight angle
- at least two obtuse angles.

2 Mark times on the clocks and draw an arrow to show:

a   a right angle

b   an acute angle

c   an obtuse angle.

3 At what time have both hands completed a revolution? _____

4 At what time can you see a straight angle? _____

Practice

1 Complete the symmetrical patterns by reflecting across the line of symmetry.

a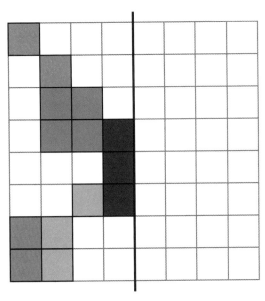

b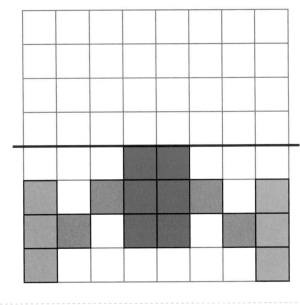

2 Identify whether each pattern is made by reflecting, rotating or translating.

a _____

b _____

c _____

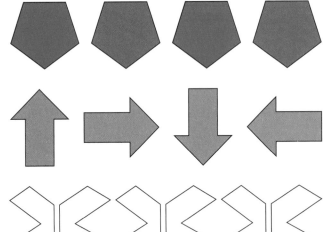

3 Which of the patterns in question 2 is not symmetrical? _____

4 Describe how each pattern was made.

a

b

_____   _____

Oxford University Press

Challenge

1 Reflect these patterns across two lines of symmetry.

a

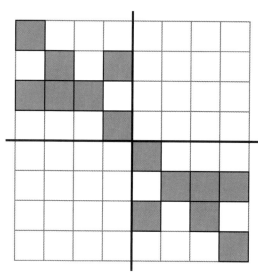

b

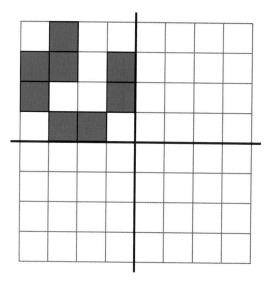

2 Use the individual shapes to create a tessellating pattern.

a

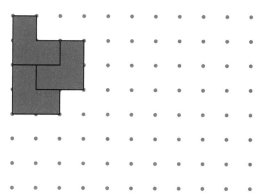

b

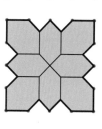

3 Reflect the shapes to create a tessellating pattern.

a

b

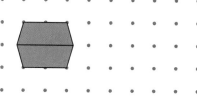

1 Draw your own symmetrical pattern on each grid. Make it reflect across both lines of symmetry.

a

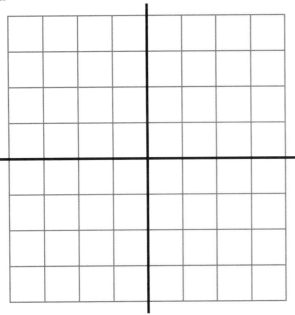

b

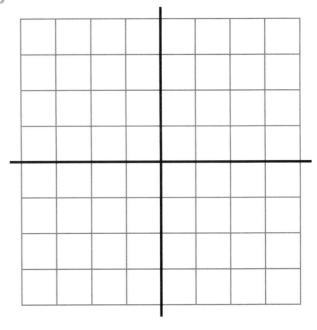

2 Make your own tessellating pattern using two different shapes for each.

a

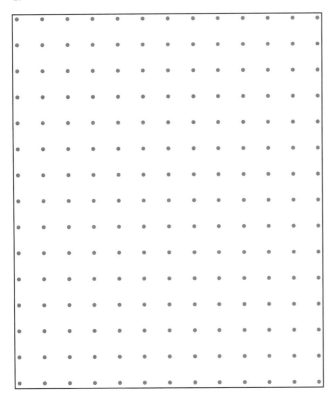

b

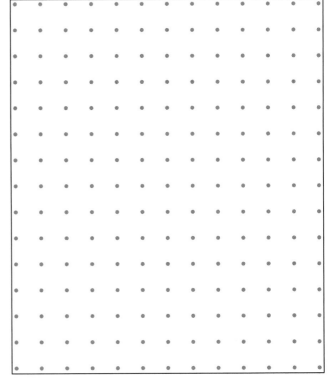

Oxford University Press

Practice

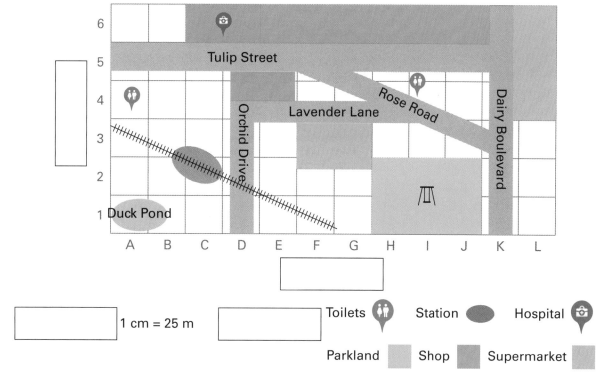

1 cm = 25 m

Toilets   Station   Hospital

Parkland   Shop   Supermarket

1 Label the following features on the map.

a   scale   b   legend   c   *x*-axis   d   *y*-axis

2 Draw the symbol used on the map to show the following.

a   the railway [   ]   b   the station [   ]   c   the playground [   ]

3 About how many metres long is:

a   Lavender Lane _____   b   Rose Road _____

c   the hospital _____   d   the duck pond? _____

4 List all the grid squares in which the following features can be found.

a   the duck pond _____

b   the park _____

c   the hospital _____

d   the station _____

e   the supermarket on Lavender Lane _____

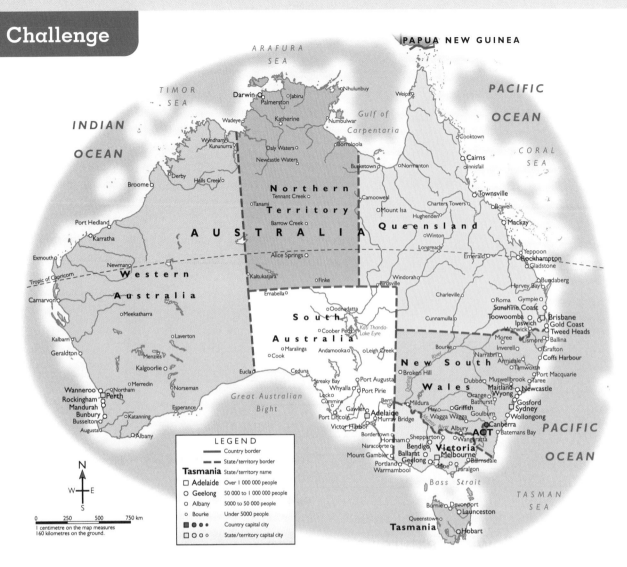

1 Use the map to find the approximate distance between:

a Perth and Halls Creek (WA) _____

b Roma and Gympie (Qld) _____

c Finke (NT) and Broken Hill (NSW) _____

d Adelaide and Sydney. _____

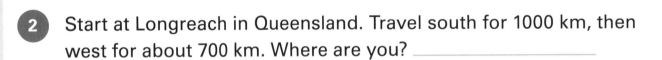

What symbols are used to show the size of cities and towns?

2 Start at Longreach in Queensland. Travel south for 1000 km, then west for about 700 km. Where are you? _____

3 Use the legend to find:

a the capital of the Northern Territory _____

b two cities with populations of over 1 million people

_____ _____

c two towns with populations of under 5000 people.

_____ _____

Oxford University Press

1. Construct your own grid map of your classroom or an area of your school. Include:

   a   a scale   b   a legend   c   a compass.

2. Write instructions for someone to move from one place to another on your map. Use as many different kinds of instructions as you can, including compass directions and information about how far needs to be travelled before each turn.

Practice

1

a   Write a survey question to ask people about their pets.

_____

_____

_____

b   List the answers you think you might get if you ask people in your class your question.

_____

_____

_____

_____

c   Conduct your survey with 10 people and record their responses below.

_____

_____

_____

_____

_____

d   Write two statements about the data you collected.

_____

_____

_____

e   Do you think your survey question was a good one?
Why or why not?

_____

_____

Oxford University Press

Challenge

1 Seth conducted a survey of 12 students. Their responses were:
red, blue, green, green, red, blue, red, blue, red, red, blue, red.

a What might his question have been? Record two possibilities.

_____

_____

b Record the results in a table.

c Show the results using tally marks.

2 Carlin surveyed 20 students to find out how many people are in their families. The most common answer was four and the highest number was six. Draw a graph showing what her data might have looked like.

1

a   Choose and record a suitable topic for a survey that would allow you to graph the results. _____

b   Write a survey question for your topic.

_____

c   List the options for answers that you could give people.

_____

d   Design a way to record your survey. You might choose to discuss with your teacher how a computer could be used for this.

e   Conduct the survey with a suitable number of people and record the results on the recording sheet.

f   Create a graph to display your results.

   Oxford University Press

Practice

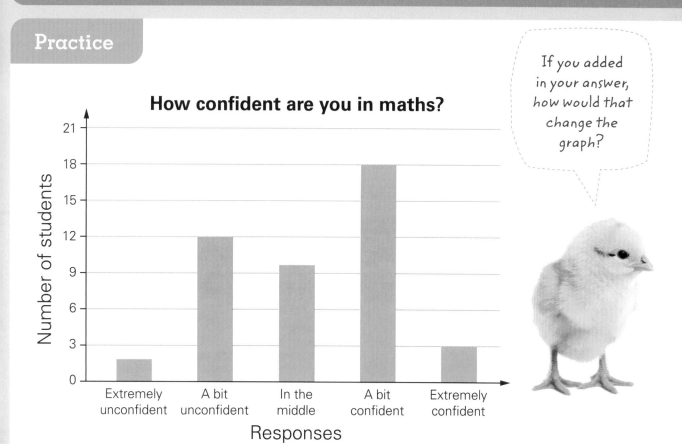

How confident are you in maths?

If you added in your answer, how would that change the graph?

1   The graph above shows how confident some Year 4 students feel about maths.

a   Make a pictograph using the same data, where one symbol represents more than one person.

b   Write three questions that can be answered by the data.

_____

_____

_____

1 These graphs show the results of a survey conducted by Year 4 students about their favourite fruits.

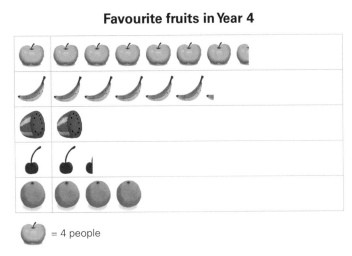

Favourite fruits in Year 4

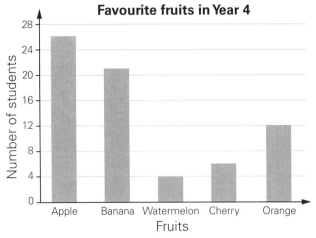

Favourite fruits in Year 4

= 4 people

a Complete the Venn diagram comparing the two ways to display the data. Think about the information that is shown on each graph.

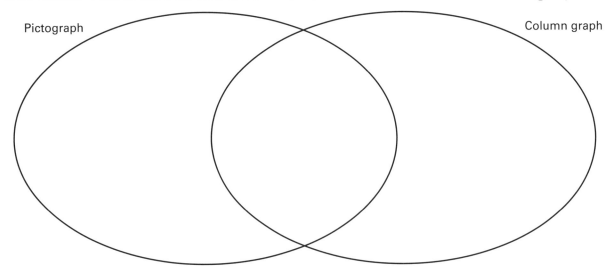

Pictograph   Column graph

b Record the results in a table.

c Which graph type do you think makes it easier to see how many people liked bananas? Why? _____

d Which graph type do you think makes it easier to see which fruit is the least popular? Why? _____

Oxford University Press

1

a   The teacher of a Year 4 class surveyed her students and found that the most popular sport was basketball. Next came football, followed by netball, soccer and cricket. Make a graph that shows what the results might look like. A computer could be a useful tool for this activity.

b   Write and answer two questions about your data.

2   In a Year 4 class, 23 out of 26 students answered *yes* to a survey question. What might the question have been? List two options.

Practice

1 List three events after each sentence.

a   It is impossible that these events will happen this week.

_____

b   It is unlikely that these events will happen this week.

_____

c   It is likely that these events will happen this week.

_____

2 Decide whether each event will affect the results in the situations described.

a   Will rolling a 3 on a dice change the likelihood of rolling a 3 on the next turn? _____

b   Will having a baby boy affect whether the next baby is a boy or a girl? _____

c   Will eating a red lolly from a bag of coloured jelly beans affect whether the next lolly you pull out will be red? _____

d   Will taking an umbrella with you affect whether it rains today?

_____

3 If it is winter in Australia today, it can't be the month of January. Write two statements about other events that can't happen at the same time.

_____

_____

_____

_____

_____

Oxford University Press

Challenge

1 Choose an appropriate term to describe the likelihood of each event happening.

a Liam lives in Central Australia. What is the likelihood of him going surfing tonight?

_____

b Faith bought 10 green apples and 30 red apples. What is the likelihood of her drawing a red apple from the bag?

_____

c Mia buys a box of chocolates. Half of them are dark chocolate and half are milk chocolate. What is the likelihood that she will choose a dark chocolate without looking?

_____

d Gracie's favourite food is pasta. How likely is she to choose spaghetti when her mum asks what she wants for dinner?

_____

e Clem is in Queensland. How likely is it that he will attend school in France today?

_____

2 Place the events from question 1 on the likelihood scale by writing in the letter for each one.

```
├─────────────────────────────────────────────┤
Impossible                                Certain
```

3 Look at any gaps on your scale. Describe two events next to *f* and *g* below that will fit in the gaps, and add them to your scale.

f _____

g _____

Imagine your class is going on an excursion.

1 Think about the probability of going to 10 different places and list them under headings such as *impossible* and *likely*. Try to include places with a range of probabilities.

2 Design a likelihood scale to show the probability of you going to each place on your list.

3 Write three statements explaining some of your choices.

Oxford University Press

Practice

1 Laney had three kinds of pasta (bowties, spaghetti and penne) and three sauces (tomato sauce, cheese sauce and pesto).

a Show all the possible meals she could make using one sauce and one pasta at a time.

b If Laney had an additional option of adding cheese, how many different meals could she make? _____

- -

2 Describe how likely each spinner is to land on red.

a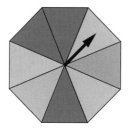

b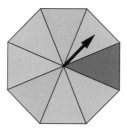

c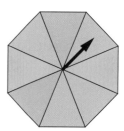

_____   _____   _____

d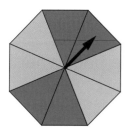

e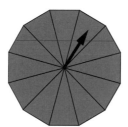

f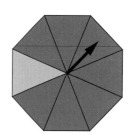

_____   _____   _____

1 Carson conducted an experiment to see which colour lolly might be drawn the most often from a bag. He completed 12 trials, returning the lollies to the bag each time. Here are his results.

Trial	1	2	3	4	5	6	7	8	9	10	11	12
Colour	blue	blue	red	blue	blue	blue	red	green	blue	blue	red	blue

    a Which colour do you think there is least of in the bag?

    _____

    b How many of each colour might there be? Write a short explanation of your answer.

    _____

    _____

    _____

    c What colour do you think Carson might draw out next? Why?

    _____

    _____

    _____

2 Amy put 10 red, 10 green and 10 yellow counters in a bag.

    a Predict the colours of the first four counters she draws out.

    _____

    b Amy draws out four yellow counters in a row. Is this the result you expected? Why or why not?

    _____

    _____

    c If Amy leaves the four yellow counters out of the bag, what is the likelihood she will draw out yellow next?

    _____

    _____

Oxford University Press

1

a   Design a chance experiment with at least three colours or items where yellow is the most likely result. Describe your experiment.

_____

_____

_____

b   List all the possible outcomes for your experiment.

c   Describe how likely each outcome is.

d   Conduct 20 trials and record the results.

e   Were the results as you expected? Write three statements describing the outcomes.

_____

_____

_____

ANSWERS

UNIT 1: Topic 1

Practice

1 a sixty thousand and ninety

   b forty-eight thousand, seven hundred and twenty-three

   c eighty-one thousand and four

2 a 38 407   b 72 095

   c 56 549

3 a 27 652 = 20 000 + 7000 + 600 + 50 + 2

   b 48 075 = 40 000 + 8000 + 70 + 5

   c 73 004 = 70 000 + 3000 + 4

   d 10 899 = 10 000 + 800 + 90 + 9

4 a 43 679   b 80 570

   c 10 005   d 64 646

Challenge

1 a 3 thousands

   b 6 hundreds

   c 9 ten thousands

   d 7000   e 90

   f 10 000

2 a 6   b 1000

   c 70 000   d 3000

   e 10   f 500

3 a 3   b 9

   c 6   d 4

   e 4   f 6

Mastery

1 Students may include a range of responses that demonstrate flexible use of place value. Likely responses include:

   40 000 + 3000 + 800 + 70 + 5

   43 000 + 800 + 70 + 5

   40 000 + 3800 + 70 + 5

   40 000 + 3000 + 870 + 5

   40 000 + 3000 + 800 + 75

   43 800 + 70 + 5

   43 870 + 5

   40 000 + 3875

   40 000 + 3000 + 875

   40 000 + 3800 + 75, etc.

2 a A range of answers is possible. Look for students

with a systematic approach to choosing numbers, e.g.

93 310, 93 301, 93 130, 93 103, 93 031, 93 013, 91 330, 91 303, 91 033

   b 93 031   93 013

   c 93 013

3 a 37 489   b 10 502

UNIT 1: Topic 2

Practice

1 a even   b odd

   c even   d even

   e odd   f even

   g even   h odd

2 a 519, 537, odd

   b 10 157, 4677, odd + even = odd

   c 7670, 6968, odd + odd = even

   d 326, 266, even − even = even

   e 1417, 3815, odd − even = odd

   f 2373, 8441, even − odd = odd

   g 24, 30, even × odd = even

   h 75, 77, odd × odd = odd

Challenge

1 a 14 262, even   b 727, odd

   c 5070, even   d 10 932, even

   e 6437, odd   f 3600, even

2 a 2224, even − even = even

   b 851, odd × odd = odd

   c $5897, odd + even = odd

Mastery

1 Multiple answers possible. End digit must be 7 or 3.

2 Multiple answers possible. Example given for each problem.

   a 517 + 432 = 949

   b 8634 + 2125 = 10 759

   c 8645 − 5402 = 3243

   d 23 × 61 = 1403

3 Multiple answers possible. Example given for each problem.

   a 4621 + 4323 = 8944

   b 6418 − 2134 = 4284

   c 5460 − 3732 = 1728

   d 54 × 35 = 1890

UNIT 1: Topic 3

Practice

1 a jump strategy

   b rearranging numbers

   c split strategy

   d rearranging numbers

   e split strategy

2 Strategies are students' own answers. You may like to share, as a class, the strategies chosen and the reasons students chose them.

   a 687   b 1878

   c 1189   d 7966

   e 1638   f 740   g 610

Challenge

1 a Sakura and Josh

   b Georgia and Cara

   c James and Fletcher

   d Jiang and Evie

   e James and Georgia

   f Sakura and Jiang

   g Josh and Cara

   h Sakura and Fletcher

2 a Sakura, Josh and Evie

   b Josh, Evie and Jiang

   c Sakura, Josh and Fletcher

   d James, Fletcher and Cara

   e James, Sakura and Fletcher

3 a Jiang and Cara (250 + 200 = 450)

   b Georgia and Josh (620 + 80 = 700)

   c Sakura and Georgia (420 + 620 = 1040)

Mastery

1 Multiple answers possible. Sample answers are:

   34 + 69 = 103   36 + 94 = 130

   34 + 96 = 130   63 + 94 = 157

   43 + 69 = 112   49 + 63 = 112

2 Multiple answers possible. Sample answers are:

   125 + 578 = 703   152 + 578 = 730

   215 + 758 = 973   155 + 728 = 883

   558 + 172 = 730   752 + 185 = 937

Oxford University Press

3 Students' own answers. Look for arrangement of digits to facilitate easy calculations and accurate calculation of the answer to students' problem.

UNIT 1: Topic 4

Practice

1 a 3846 + 2523
= 6 + 3 + 40 + 20 + 800 + 500 + 3000 + 2000
= 9 + 60 + 1300 + 5000
= 6369

b 14254 + 32817
= 4 + 7 + 50 + 10 + 200 + 800 + 4000 + 2000 + 10000 + 30000
= 11 + 60 + 1000 + 6000 + 40000
= 47071

c 59346 + 8282
= 6 + 2 + 40 + 80 + 300 + 200 + 9000 + 8000 + 50000
= 8 + 120 + 500 + 17000 + 50000
= 67628

d 3794 +26436
= 4 + 6 + 90 + 30 + 700 + 400 + 3000 + 6000 + 20000
= 10 + 120 + 1100 + 9000 + 20000
= 30230

Challenge

1 a 80299
b 80302
c Isla and Maya's team

2 a 85473   b 85470
c Mason   d Owen

Mastery

1 a Multiple answers possible. Sample answers are:
47826 + 43145 = 90971
67816 + 23155 = 90971

b Multiple answers possible. Sample answers are:
32607 + 51585 = 84192
32207 + 51985 = 84192

c Multiple answers possible. Sample answers are:
47774 + 35525 = 83299
47674 + 35625 = 83299

UNIT 1: Topic 5

Practice

1 a 65

b 196

c 461
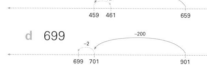

d 699

2 Strategies best suited to the problem are listed, however, if students successfully use the other strategy, mark their choice as correct.

a 7   Adding on
b 334   Compensation
c 1358   Compensation
d 12   Either
e 5   Either
f 2778   Compensation

Challenge

1 a 306   b 16
c 3B and 4D   d 233
e 834

2 Estimated answers may vary depending on the number students round to. Most likely answers are given.

a Estimated answer: 323
Exact answer: 324

b Estimated answer: 3027
Exact answer: 3029

c Estimated answer: 608
Exact answer: 609

d Estimated answer: 4831
Exact answer: 4829

e Estimated answer: 485
Exact answer: 488

f Estimated answer: 3140
Exact answer: 3139

g Estimated answer: 462
Exact answer: 464

Mastery

1 Multiple answers possible. Look for subtrahends that are close to 150. Sample answers are:
732 – 149, 732 – 151

2 Multiple answers possible. Look for subtrahends that are close to 800. Sample answers are:
3184 – 799, 3184 – 801

3 Multiple answers possible. Look for numbers that are easily rounded to the nearest 10 or 100. Sample answers are:
651 – 398 = 253, 1368 – 59 = 1309

UNIT 1: Topic 6

Practice

1 a 7526 – 3214 =
7526 – 4 = 7522
– 10 = 7512 – 200 = 7312
– 3000 = 4312

b 8140 – 5624 =
8140 – 4 = 8136
– 20 = 8116 – 600 = 7516
– 5000 = 2516

c 29107 – 17804 =
29107 – 4 = 29103
– 800 = 28303
– 7000 = 21303
– 10000 = 11303

d 47264 – 24682 =
47264 – 2 = 47262
– 80 = 47182
– 600 = 46582
– 4000 = 42582
– 20000 = 22582

ANSWERS

Challenge

1 **a** $15054   **b** 4170 km

2 **a**

Month	Amount	Amount left to raise
January	$4725	$95274
February	$11092	$84182
March	$8634	$75548
April	$23905	$51643
May	$19712	$31931
June	$25560	$6371

  b $25560 − $4725 = $20835

  c $25560 − $23905 = $1655

Mastery

1 **a** Multiple answers possible.
Sample answers are:
75387 − 34182 = 41205,
55307 − 14102 = 41205

  b Multiple answers possible.
Sample answers are:
92678 − 28306 = 64372,
91675 − 27303 = 64372

  c Multiple answers possible.
Sample answers are:
73984 − 31486 = 42498,
73784 − 31286 = 42498

2 **a & b** Multiple answers possible.
Look for students who are able
to flexibly use subtraction to find
answers that meet the criteria,
and who use patterns to generate
different options.

UNIT 1: Topic 7

Practice

1 **a** 20 ÷ 4 = 5   20 ÷ 5 = 4

  b 42 ÷ 6 = 7   42 ÷ 7 = 6

  c 72 ÷ 9 = 8   72 ÷ 8 = 9

  d 21 ÷ 3 = 7   21 ÷ 7 = 3

  e 54 ÷ 6 = 9   54 ÷ 9 = 6

2 **a** 10 × 2 = 20   2 × 10 = 20

  b 8 × 7 = 56   7 × 8 = 56

  c 6 × 8 = 48   8 × 6 = 48

  d 4 × 9 = 36   9 × 4 = 36

  e 10 × 5 = 50   5 × 10 = 50

3 **a** 9 × 7 = 63   7 × 9 = 63
    63 ÷ 9 = 7   63 ÷ 7 = 9

  b 4 × 10 = 40   10 × 4 = 40
    40 ÷ 10 = 4   40 ÷ 4 = 10

Challenge

1 **a** 3 × 9 = 27, 9 × 3 = 27,
27 ÷ 3 = 9, 27 ÷ 9 = 3

  b 4 × 8 = 32, 8 × 4 = 32,
32 ÷ 4 = 8, 32 ÷ 8 = 4

  c 10 × 7 = 70, 7 × 10 = 70,
70 ÷ 10 = 7, 70 ÷ 7 = 10

  d 2 × 9 = 18, 9 × 2 = 18,
18 ÷ 2 = 9, 18 ÷ 9 = 2

  e 9 × 8 = 72, 8 × 9 = 72,
72 ÷ 9 = 8, 72 ÷ 8 = 9

2 **a** 36   **b** 14, 28   **c** 16, 32

3 **a** 9   **b** 10   **c** 7

  d 8   **e** 5   **f** 10

  g 35   **h** 28   **i** 81

  j 100   **k** 3   **l** 7

4 **a** Teacher to check. Possible
answer: They are all square
numbers.

  b 36 and 49

Mastery

1 Possible answers within
10 × 10 times tables are as
follows. Students may also come
up with larger answers.

1 × 9 = 9, 9 × 1 = 9, 3 × 3 = 9

9 ÷ 1 = 9, 18 ÷ 2 = 9, 27 ÷ 3 = 9,
36 ÷ 4 = 9, 45 ÷ 5 = 9, 54 ÷ 6 = 9,
63 ÷ 7 = 9, 72 ÷ 8 = 9, 81 ÷ 9 = 9,
90 ÷ 10 = 9

2 24 ÷ 24 = 1, 24 ÷ 2 = 12,
24 ÷ 12 = 2, 24 ÷ 3 = 8,
24 ÷ 8 = 3, 24 ÷ 4 = 6, 24 ÷ 6 = 4

3 **a** 8 × 7 = 56

  b 63 ÷ 7 = 9

  c 150 ÷ 10 = 15

  d 5 × $7 = $35

  e 6 × 4 = 24

UNIT 1: Topic 8

Practice

1 **a** Extended   Contracted

  b Extended   Contracted

  c Extended   Contracted

  d Extended   Contracted

  e Extended   Contracted

  f Extended   Contracted

Challenge

1 **a**   **b**

  c   **d**

2 **a** 236 cm   **b** 354 cm

  c 413 cm   **d** 531 cm

Oxford University Press

Mastery

1 Possible combinations are:

$3 \times 68 = 204$ $3 \times 86 = 258$
$6 \times 38 = 228$ $6 \times 83 = 498$
$8 \times 36 = 288$ $8 \times 63 = 504$

2 Multiple combinations possible. Look for students who can correctly lay out problems using their chosen method and who can accurately calculate the answers.

UNIT 1: Topic 9

Practice

1 a
```
      2 8
   3 ) 8 4
```
	T	O
	2	8
×		3
	8	4

b
```
      1 9
   5 ) 9 5
```
	T	O
	1	9
×		5
	9	5

c
```
      1 1
   9 ) 9 9
```
	T	O
	1	1
×		9
	9	9

2 a 52   b 27
   c 24   d 41
   e 68   f 27

Challenge

1 a Yes   b Yes   c No
2 a Yes   b Yes   c No
3 a
```
      3 6
   4 ) 1 4 4
```
b
```
      1 0 3
   3 ) 3 0 9
```
c
```
      3 9 9
   2 ) 7 8 9
```

Mastery

1 162, 171, 180, 189, 198, 207, 216, 225, 234, 243, 252

2 189, 252

3 a Possible equations are:
   $428 \div 2 = 214$, $448 \div 2 = 224$,
   $468 \div 2 = 234$, $488 \div 2 = 244$

   b Possible equations are:
   $432 \div 3 = 144$, $435 \div 3 = 145$,
   $438 \div 3 = 146$

   c Possible equations are:
   $736 \div 4 = 184$, $776 \div 4 = 194$

UNIT 2: Topic 1

Practice

1 a 8   b 2   c 6
   d 9   e 6   f 2

2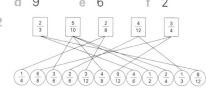

Challenge

1 a not equivalent
   b equivalent
   c equivalent
   d not equivalent

2 a Zan
   b Zan
   c Dan
   d Zan

3 Teacher to check drawings. Likely equivalents are:
   a $\frac{8}{10}$
   b $\frac{3}{4}$ or $\frac{6}{8}$

4 Teacher to check drawings. Likely equivalents are:
   a $\frac{2}{4}, \frac{3}{6}, \frac{4}{8}, \frac{5}{10}$ or $\frac{6}{12}$
   b $\frac{2}{3}$ and $\frac{8}{12}$

Mastery

1 Herbie $\frac{1}{2}$, Carly $\frac{1}{4}$ and Erwin $\frac{2}{8}$

2 Theo $\frac{1}{3}$, Sam $\frac{2}{6}$ and Toni $\frac{4}{12}$

   Theo $\frac{1}{3}$, Sam $\frac{1}{6}$ and Toni $\frac{6}{12}$

   Theo $\frac{1}{3}$, Sam $\frac{3}{6}$ and Toni $\frac{2}{12}$

   Theo $\frac{2}{3}$, Sam $\frac{1}{6}$ and Toni $\frac{2}{12}$

3 Multiple answers possible, e.g.
$\frac{1}{5}$ and $\frac{3}{10}$; $\frac{2}{5}$ and $\frac{1}{10}$; $\frac{1}{3}$ and $\frac{1}{6}$; $\frac{1}{4}$ and $\frac{2}{8}$; $\frac{1}{4}$ and $\frac{3}{12}$; $\frac{1}{6}$ and $\frac{4}{12}$, etc.

UNIT 2: Topic 2

Practice

1
Diagram	Improper fraction	Mixed number
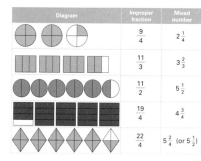	$\frac{9}{4}$	$2\frac{1}{4}$
	$\frac{11}{3}$	$3\frac{2}{3}$
	$\frac{11}{2}$	$5\frac{1}{2}$
	$\frac{19}{4}$	$4\frac{3}{4}$
	$\frac{22}{4}$	$5\frac{2}{4}$ (or $5\frac{1}{2}$)

2 a 4

   b $4\frac{1}{2}$

   c $3\frac{3}{4}$

3 a $\frac{11}{2}$
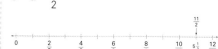

   b $\frac{9}{4}$

   c $\frac{14}{3}$

Challenge

1 Teacher to check diagrams.
   a $\frac{7}{2} = 3\frac{1}{2}$ b $\frac{10}{3} = 3\frac{1}{3}$
   c $\frac{5}{3} = 1\frac{2}{3}$ d $\frac{13}{4} = 3\frac{1}{4}$

2 a 1, $1\frac{1}{3}$, $1\frac{2}{3}$, 2, $2\frac{1}{3}$, $2\frac{2}{3}$, 3, $3\frac{1}{3}$, $3\frac{2}{3}$, 4

   b $\frac{3}{3}, \frac{4}{3}, \frac{5}{3}, \frac{6}{3}, \frac{7}{3}, \frac{8}{3}, \frac{9}{3}, \frac{10}{3}, \frac{11}{3}, \frac{12}{3}$

3 a $\frac{38}{4}$, $9\frac{2}{4}$ or $9\frac{1}{2}$ b $\frac{13}{3}$, $4\frac{1}{3}$

   c $\frac{10}{2}$, 5

ANSWERS

Mastery

1   Labels on number line should be:
Mixed numbers: 1, $1\frac{1}{8}$, $1\frac{2}{8}$, $1\frac{3}{8}$,

$1\frac{4}{8}$, $1\frac{5}{8}$, $1\frac{6}{8}$, $1\frac{7}{8}$, 2, $2\frac{1}{8}$, $2\frac{2}{8}$, $2\frac{3}{8}$,

$2\frac{4}{8}$, $2\frac{5}{8}$, $2\frac{6}{8}$, $2\frac{7}{8}$, 3

Fractions: $\frac{8}{8}$, $\frac{9}{8}$, $\frac{10}{8}$, $\frac{11}{8}$, $\frac{12}{8}$, $\frac{13}{8}$, $\frac{14}{8}$,

$\frac{15}{8}$, $\frac{16}{8}$, $\frac{17}{8}$, $\frac{18}{8}$, $\frac{19}{8}$, $\frac{20}{8}$, $\frac{21}{8}$, $\frac{22}{8}$, $\frac{23}{8}$, $\frac{24}{8}$

2   a   Multiple answers possible,
e.g. $1\frac{3}{8}$, $2\frac{3}{8}$ and $3\frac{3}{8}$.

b   Number lines will vary
depending on the mixed
numbers chosen by students.

3   a   Multiple answers possible,
e.g. $1\frac{5}{6}$, $2\frac{5}{6}$ and $3\frac{5}{6}$.

b   Number lines will vary
depending on the mixed
numbers chosen by students.

4   $\frac{30}{5}$, $\frac{31}{5}$, $\frac{32}{5}$, $\frac{33}{5}$, $\frac{34}{5}$, $\frac{35}{5}$, $\frac{36}{5}$, $\frac{37}{5}$, $\frac{38}{5}$,

$\frac{39}{5}$, $\frac{40}{5}$

UNIT 2: Topic 3

Practice

1   a

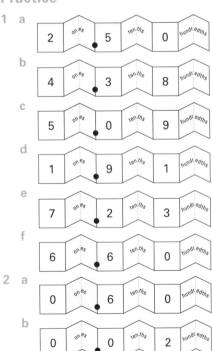

2   a

c

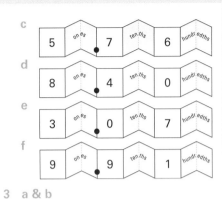

3   a & b

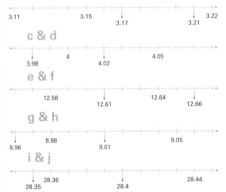

c & d

e & f

g & h

i & j

Challenge

1   a   7.1   b   7.8

c   4.95   d   5.02

e   5.1   f   15.46

g   15.51   h   15.6

2   a   Kimiko

b   Harley and Toby

c   8.4   d $8\frac{88}{100}$ e $\frac{64}{100}$ or 0.64

3   a $49\frac{99}{100}$ b   49.09   c   50.8

d   49.09, $50\frac{3}{100}$ or 50.03

Mastery

1   a   Multiple answers possible.
Sample answers are:

36.24, 42.36, 23.64, 62.34, etc.

b & c   Answers will depend on
students' responses to a.

2   Multiple answers possible, e.g.
23.54, 23.62, 23.75, 23.81, 24.91,
24.98

3   a & b   Teacher to check. Look for
answers that show students can
use different combinations
of decimal numbers and convert
from common fractions to
decimals.

UNIT 3: Topic 1

Practice

1   a $8.80   b $56.25

c $89.85   d $1.00

e $20.00   f $132.55

g $518.60   h $3.00

2

Amount	Rounds to	Change from $10	Change from $20	Change from $50
$4.32	$4.30	$5.70	$15.70	$45.70
$0.06	$0.05	$9.95	$19.95	$49.95
$3.74	$3.75	$6.25	$16.25	$46.25
$4.98	$5.00	$5.00	$15.00	$45.00
$1.17	$1.15	$8.85	$18.85	$48.85

Challenge

1   a   down

b $23.20 if using rounded
numbers ($23.19 if using
unrounded totals)

c   the book and the cake

d $34.56

e   the shirt and the shoes

f $41.60

2   a $132   b $105

c   No

d   Answers may vary. Most likely
answer is that Amber earned
$50 or $100.

e   Answers will depend on
response to previous question
e.g. if Amber earned $50, the
gifts would have cost $21.55
and if she earned $100, the
gifts would have cost $71.55.

Mastery

1   a–c   Multiple answers possible,
e.g. Item A and Item B total $8.89
and would round up to $8.90.
Change of $41.10 would be
received from $50.

2   Multiple answers possible e.g.
if Item 1 cost $3.20, Item 2 cost
$4.89 and Item 3 cost $4.39, the
total cost would be $12.48, which
would round up to $12.50.

3   Multiple answers possible e.g. if
Amelia paid with $20, the total
cost would have been $12.80. One
item might therefore have cost
$8.76 and the other $4.04.

Oxford University Press

UNIT 4: Topic 1

Practice

1 a 1 2 4 8 16 32 64 128 256 512 1024

   Rule: × 2

   b 3 9 15 21 27 33 39 45 51 57

   Rule: + 6

   c 152 148 144 140 136 132 128 124 120 116

   Rule: − 4

2 a, b & d

	1	2	3	4	5	6	7	8	9
0	1	2	3	4	5	6	7	8	9
10	11	12	13	14	15	16	17	18	19
20	21	22	23	24	25	26	27	28	29
30	31	32	33	34	35	36	37	38	39
40	41	42	43	44	45	46	47	48	49
50	51	52	53	54	55	56	57	58	59
60	61	62	63	64	65	66	67	68	69
70	71	72	73	74	75	76	77	78	79
80	81	82	83	84	85	86	87	88	89
90	91	92	93	94	95	96	97	98	99

   c 0, 6, 12, 18, 24, 30, 36, 42, 48, 54, 60, 66, 72, 78, 84, 90, 96

   e 6

   f 2, 1, 0, 9, 8, 7, 6, 5, 4, 3

Challenge

1 a

3	11	19	27	35	43
51	**59**	67	75	83	91
99	107	115	**123**	131	139
147	155	**163**	171	179	**187**
195	203	211	**219**	227	235
243	251	**259**	267	275	**283**

   Rule: add 8

   b

253	246	239	232	**225**	218
211	**204**	197	190	183	176
169	162	155	**148**	141	134
127	120	113	106	99	92
85	78	71	64	**57**	50
43	**36**	29	22	15	8

   Rule: take away 7

2 a 3

   b 9

   c 27

d Answers may vary. Likely response: The calculator multiplies the previous answer by three.

3 a 49, 42, 35, 63

   b 56, 96, 32, 88

   c 27, 72, 81, 63, 45

Mastery

1 Answers will vary, e.g. 1 and 6, 2 and 12, 3 and 18, 4 and 24, 5 and 30

2 a multiply by 3 or add 8

   b multiply by 3

   c Answers will vary e.g. 1 and 3, 2 and 6, 3 and 9, 5 and 15, 6 and 18

3 a Answers will vary, e.g. multiply by 10, add 5, take away 10, etc.

   b Teacher to check. Answer will depend on rule chosen by student.

   c Teacher to check. Answer will depend on rule chosen by student.

UNIT 4: Topic 2

Practice

1 $10 \times 2 = 48 - 28$
$52 + 21 = 28 + 45$
$76 - 52 = 5 + 19$
$63 \div 7 = 38 - 29$
$6 \times 9 = 38 + 16$

2 a 9   b 44

   c 54   d 30

   e 100   f 8

   g 36   g 5

3 The following equations should be circled: $100 - 15$, $47 + 38$, $170 \div 2$, $109 - 24$, 17×5, $71 + 14$

Challenge

1 a $58 - 17 = 18 + 23$

   b $27 \div 3 = 48 - 39$

   c $32 + 24 = 8 \times 7$

   d $95 - 44 = 3 + 48$

   e $12 \times 4 = 6 \times 8$

   f $35 + 57 = 99 - 7$

   g $86 - 40 = 78 - 32$

   h $8 \times 4 = 63 - 31$

2 a $87 - 28 = 31 + 28$, Linus subtracted 28.

   b $24 + 42 = 33 \times 2$, Hala started with 42.

   c $3 \times 15 = 92 - 47$, Jacob started with 15.

   d $82 - 21 = 34 + 27$, Andy started with 82.

Mastery

1 Multiple answers possible. Sample answers are:

   $22 + 34 = 99 - 43$, $23 + 33 = 89 - 33$, $29 + 37 = 99 - 33$

2 Multiple answers possible. Sample answers are:

   $100 \div 2 = 28 + 22$, $86 \div 2 = 20 + 23$, $16 \div 2 = 5 + 3$

3 Multiple answers possible. Sample answers are:

   $21 + 21 + 32 = 74$, $10 + 10 + 32 = 52$, $35 + 35 + 32 = 102$

UNIT 5: Topic 1

Practice

1 Teacher to check estimates.

   a 30 mm   b 50 mm

   c 35 mm   d 38 mm

   e 8 mm   f 32 mm

2 a 8 cm   b 11.5 cm

   c 9.5 cm   d 4 cm

   e 6 cm

3 a 30 cm tie

   b $17\frac{1}{2}$ cm worm

   c 25 cm book

   d 2 m rope

Challenge

1 a 13.5 cm   b 3 cm

   c 17.5 cm

2 a 175 mm   b 287 mm

3 3 cm, 13.5 cm, 17.5 cm/175 mm, 287 mm

4 a 38.5 cm or 385 mm

   b 10.5 cm or 105 mm

   c 92 cm or 920 mm

   d 4.5 m or 450 cm

ANSWERS

Mastery

1. **a & b** Teacher to check drawings.
2. Multiple answers possible, e.g. five sides of 13 cm each or three sides of 15 cm and two of 10 cm.
3. Multiple answers possible. Teacher to check.

UNIT 5: Topic 2

Practice

1. a m² b m²
   c cm² d cm²
   e m² f cm²
2. a 20 cm² b 19 cm²
   c 9 cm² d 12 cm²

Challenge

1. **a–f** Teacher to check
2. a a b 23 cm²

Mastery

1. Teacher to check
2. a 9 m² b 20 m²
   c 14 m² d 24 m²
   e 18 m² f 36 m²

UNIT 5: Topic 3

Practice

1. a 8 cm³ b 11 cm³
   c 14 cm³
2. **a & b** Answers will vary. Teacher to check.
3.

Challenge

1. a Teacher to check b 18 cm³
2. a Teacher to check b 16 cm³

---

3. a 2500 mL, 2 L 500mL
   b 400 mL, 0.4 L
   c 1100 mL, 1.1 L
   d 3000 mL, 3 L
   e 4600 mL, 4.6 L
   f 470 mL, 0.47 L

Mastery

1. Teacher to check. Possibilities include eight layers of 3, six layers of 4, 2 layers of 12, and the reverse of each of these.
2. Teacher to check. Look for students who can accurately space the markings and understand how increments work.

UNIT 5: Topic 4

Practice

1. C, E, B, A, D
2. a Each banana has a mass of 120 g.
   b Each rabbit has a mass of 1.5 kg.
   c The watermelon has a mass of 1 kg.
   d Each bunch of grapes has a mass of 300 g. Each packet of chips has a mass of 100g.

Challenge

1. 535 grams
2. a

   b Angela's c 4 cups
3. Teacher to decide on an acceptable level of accuracy for showing the masses in question 3C and 3G.

Mastery

1. **a & b** Answers will vary. Look for demonstration of students' understanding of the relative masses of common items.

---

2. Teacher to check. Look for appropriate choice of scale and reasonable estimates as to the masses of students' chosen items and places.

UNIT 5: Topic 5

Practice

1. a hotter b colder
   c colder d hotter
2.

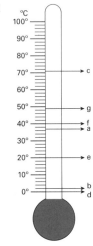

Challenge

1. a

Time	Temperature
8 am	4°C
9 am	4°C
10 am	24°C
11 am	32°C
12 pm	52°C
1 pm	57°C
2 pm	68°C
3 pm	78°C
4 pm	89°C
5 pm	89°C
6 pm	63°C
7 pm	35°C
8 pm	16°C

   b 4 pm to 5 pm
   c 8 am to 9 am
2. a Alice Springs
   b Alice Springs
   c Planet Creeb
   d 10 am
   e Students' own answers. Look for recognition that the temperature is hot in Alice Springs so it's likely to be summer and unlikely to be winter.

Oxford University Press

Mastery

1 **a & b** Answers will vary. Look for demonstrated understanding of Celsius temperatures and how they relate to the temperature during the day.

2 Teacher to check. Look for choice of an appropriate scale for students' thermometers and reasonable estimates as to the temperatures of their chosen items and places.

UNIT 5: Topic 6

Practice

1 earliest 3:14 am 4:33 am 6:54 am 3:16 pm 6:02 pm 11:38 pm latest

2 a

Departed from	Time	Arrived at	Time	Travel time
Salt City	7:43 am	Butter Bay	7:58 am	15 minutes
Butter Bay	8:16 am	Herb Hill	9:03 am	47 minutes
Herb Hill	10:30 am	Garlic Green	11:09 am	39 minutes
Garlic Green	11:14 am	Cherry Creek	11:23 am	9 minutes
Cherry Creek	11:48 am	Banana Bend	12:07 pm	19 minutes
Banana Bend	1:51 pm	Salt City	2:02 pm	11 minutes

 b 25 minutes

 c Banana Bend

 d Garlic Green

 e 6 hours and 19 minutes

 f afternoon

 g 300 seconds

Challenge

1 a 6 hours and 30 minutes

 b 5 hours

 c 25 hours

 d

 e pm

 f 4 hours and 35 minutes

---

2 a 8:23 am 8:42 am

 b 19 minutes

 c 3:46 pm

3 a about 4 months

 b 2 minutes and 35 seconds

 c school holidays

Mastery

1 Answers will vary. Look for understanding of the placement of numbers around an analog clock and accurately represented times.

2 Answers will vary. Look for accurately calculated elapsed times from a starting time.

3 **a–c** Answers will vary. Look for demonstrated understanding of duration by suggesting plausible activities for each unit of time.

UNIT 6: Topic 1

Practice

1 a hexagon

 b irregular

 c triangle, pentagon

2 a octagon

 b regular

 c trapezium, hexagon

3 a B   b D

 c E   d A

4 a 2   b 2

 c 4   d 3

Challenge

1 a Students circle the words octagon and equilateral triangle.

 b Teacher to check that the named 2D shapes match the picture. Discussion could take place beforehand about which 2D shapes might be appropriate.

2 **a–c** Teacher to check that the shapes are irregular, have been correctly named, and that their area has been accurately determined.

Mastery

1 **a–d** Answers will vary. Check for accuracy of area of each shape and of the names of the split shapes.

2 Answers will vary. Check the accuracy of the naming of each shape.

UNIT 6: Topic 2

Practice

1 The following pictures are a guide. Teacher to check students' drawings.

 a   b

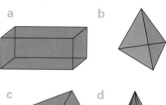

 c   d

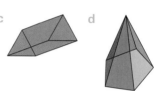

2 **a–c** Teacher to check

Challenge

1

2 a 8 faces, 18 edges, 12 corners

 b 8 faces, 18 edges, 12 corners

 c 8 faces, 18 edges, 12 corners

 d 10 faces, 24 edges, 16 corners

Mastery

1 Teacher to check. Possibilities include a cube, a rectangular prism and a trapezium prism.

2 Teacher to check. Any prism will meet the criteria, as well as a cylinder. Students may also draw an object made from cubes with a rectangular side view.

3 a rectangular prism

 b triangular prism

UNIT 7: Topic 1

Practice

1 **a–d** Teacher to check accurate comparisons of angle sizes against a right angle.

ANSWERS

2 Sample answers shown. Students may find other examples.

3 a–c Teacher to check

4 The second, third and fifth shapes should be circled.

Challenge

1 a–c Teacher to check. Examples of possible shapes are:

a square or rectangle

b right-angled triangle or irregular shapes

c equilateral, isosceles or scalene triangle, regular hexagon, regular pentagon

2 a–d Teacher to check

3 a E   b A, B, C, D

c B, C

Mastery

1 Teacher to check

2 Teacher to check

3 12 o'clock

4 Most common answer is likely to be 6 o'clock. Other possibilities are around: 12.32; 1.38; 2.43; 3.49; 4.54; 7.05; 8.10; 9.16; 10.21 & 11.27

UNIT 8: Topic 1

Practice

1 a

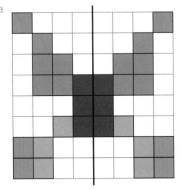

---

b

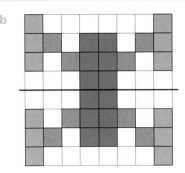

2 a translating   b rotating

c reflecting

3 pattern b

4 a Could be either rotated or reflected. Look for students who are able to use the language of transformations, such as half turns.

b Object is rotated. Look for students who offer extra information, such as the fact that the turns are quarter turns clockwise.

Challenge

1 a

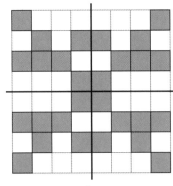

b

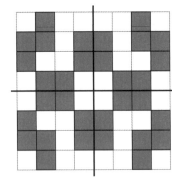

2 a Teacher to check. Possible answers are shown.

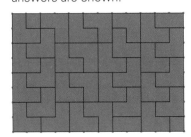

---

b

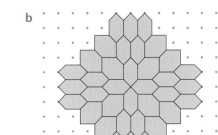

3 a

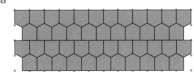

b

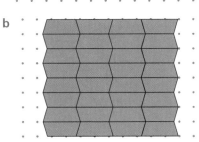

Mastery

1 a & b Teacher to check. Look for successful reflection of the pattern through all four quadrants.

2 a & b Teacher to check. Look for understanding of the concept of tessellation and application to find shapes that tessellate with other shapes.

UNIT 8: Topic 2

Practice

1 a–d

2 a ─┼┼┼┼┼┼┼─   b

c

3 Teacher: Answers below are approximate. Accept answers within a reasonable range.

   a About 125 metres

   b About 145 metres

   c About 50 metres

   d About 38 metres

4 a A1 & B1

   b H1, I1, J1, H2, I2 & J2

   c C6 & D6

   d B2, B3, C2 & C3

   e F2, F3, G2 & G3

Challenge

1 Teacher: accept answers within a reasonable range.

   a Approximately 2250 km

   b Approximately 450 km

   c Approximately 1125 km

   d Approximately 1300 km

2 Port Pirie

3 a Darwin

   b Possibilities are: Sydney, Adelaide, Melbourne, Perth and Brisbane

   c Teacher to check: multiple answers possible.

Mastery

1 a–c Teacher to check. Look for accurate representation of familiar locations and successful use of tools and language of location.

2 Teacher to check. Look for understanding of a range of vocabulary and conventions relating to location.

UNIT 9: Topic 1

Practice

1 a Teacher to check. Look for understanding that the question has limited possibilities for responses, rather than asking for opinions or other more open-ended questions.

   b Answers will depend on answer to a. Look for plausible responses to the question.

c Teacher to check. Look for appropriate way to record data, such as a list, frequency table or tally marks.

d Teacher to check. Look for accurate interpretation of the data collected and the choice of significant elements of the data on which to base observations.

e Teacher to check. Look for recognition of whether or not the data was useful and whether students' question elicited the kind of information that they anticipated.

Challenge

1 a Teacher to check. Examples are *Out of red, green and blue, which colour do you prefer?* or *What house are you in at school?*

   b Sample table below. Accept reasonable variations based upon the data.

Colour	Red	Blue	Green
Number of people	6	4	2

   c Red   ⃦⃦⃦⃦ |

       Blue ||||

       Green ||

2 Teacher to check. Look for correct interpretation of criteria and appropriate graph form to represent what the data might reasonably be.

Mastery

1 a Teacher to check. Look for appropriate topics that will allow students to collect the required data.

   b Teacher to check. Look for understanding of how to write an appropriate survey question for their chosen topic.

   c Teacher to check. Answers will depend on the question written by the student.

   d & e Teacher to check. Look for understanding of how to collect and record survey data.

   f Teacher to check. Look for appropriate forms of data display and accurate representation of collected data.

UNIT 9: Topic 2

Practice

1 a Answers will vary depending on the scale chosen by students. Sample shown below.

   b Answers will vary. Possible questions might include: *Which response had the most answers? How many people feel a bit confident or extremely confident in maths?* and *Which response did 10 students choose?*

Challenge

1 a Teacher to check. Possible observations might include that both graphs show the least and most popular fruits, the values on both graphs are the same, the column graph has numbers to show the data and the pictograph uses pictures.

   b Students' own answer. Look for an understanding of how to interpret data in the justification of students' answer.

   c & d Students' own answer. Look for an understanding of how to interpret data in the justification of students' answer.

Mastery

1 a Teacher to check. Look for an appropriate graph type for the data and plausible values for each category.

   b Students' own answers. Look for competently interpreted data through appropriate questions and accurate answers.

2 Students' own answers. Look for recognition that the question is likely to be about something that students enjoy, such as who wants to go out for sport now, or about a category that the majority of students fit into, such as who brought a bag to school today.

ANSWERS

UNIT 10: Topic 1

Practice

1 a–c Students' own answers. Look for a good understanding of the language of probability through the choice of appropriate activities.

2 a no   b no

   c yes   d no

3 Teacher to check: Look for an understanding of events that are mutually exclusive.

Challenge

1 Answers may vary – accept any reasonable response. Likely answers are:

   a very unlikely

   b likely

   c even chance

   d very likely

   e impossible

2
Impossible   Certain

3 Teacher to check. Look for the ability to choose two appropriate events to fit into the gaps on students' scales and the use of language or probability to accurately describe likelihood.

Mastery

1 Teacher to check. Look for an understanding of a range of probability terms and appropriately matched familiar places with the correct language.

2 Teacher to check. Look for accurate translations of the observations students made into a visual representation of likelihood.

3 Teacher to check. Look for use of language of probability to justify the reasons for students' placement of various items on the scale.

UNIT 10: Topic 2

Practice

1 a There are nine possible combinations:

   bowties with cheese sauce
   bowties with tomato sauce
   bowties with pesto

   spaghetti with cheese sauce
   spaghetti with tomato sauce
   spaghetti with pesto

   penne with cheese sauce
   penne with tomato sauce
   penne with pesto

   b 18 – the nine original options, each with or without cheese.

2 a–f Answers may vary. Most likely answers shown, but accept any reasonable evaluation of the probability of landing on red.

   a likely   b very unlikely

   c impossible   d likely

   e certain   f very likely

Challenge

1 a Green is the most likely answer, but accept alternatives if students can justify their responses.

   b Students' own answers. Look for students who recognise there are likely to be more blue than any other colour and who can offer justification of their response using the language of probability.

   c Students' own answers. Look for students who recognise that blue is the most likely response even though it has come up the most often and who show understanding that the previous result will not influence the next result.

2 a Students' own answers. Look for students who recognise there is an equal chance of choosing any one colour.

   b Most likely answer is no, since there are fewer yellow than red or green counters after each yellow counter has been taken out, but accept any response with a reasonable justification.

   c Exact term chosen may vary. Look for students who recognise there is now even less chance of drawing out a yellow counter because only six of the remaining 26 counters are yellow, whereas there are still 10 red and 10 green counters.

Mastery

1 a Teacher to check. Look for choice of an appropriate experiment and explanations of why yellow will be the most likely result.

   b Teacher to check. Answers will depend on students' chosen experiment. Look for systematic identification of the possible outcomes.

   c Teacher to check. Answers will depend on outcomes identified in b. Look for use of appropriate language of probability to describe the likelihood of each outcome.

   d Teacher to check. Look for choice of an appropriate method to record data, such as a list or table.

   e Teacher to check. Look for analysis of actual results in light of the probable results.

Oxford University Press

Oxford University Press